Praise for *The Hidden Spring*:

THE HIDDEN SPRING

A Journey to the Source of Consciousness

MARK SOLMS

P

PROFILE BOOKS

This paperback edition first published in 2022

First published in Great Britain in 2021 by
Profile Books Ltd
29 Cloth Fair
London
ECIA 7JQ

www.profilebooks.com

3 5 7 9 10 8 6 4 2

Typeset in Sabon by MacGuru Ltd
Printed and bound in Great Britain by
CPI Group (UK) Ltd, Croydon, CRO 4YY

A CIP catalogue record for this book is available from the British Library.

ISBN 978 1 78816 284 5
eISBN 978 1 78283 571 4
Export ISBN 978 1 78816 762 8

In memory of
Jaak Panksepp
(1943–2017)

He solved the ancient riddle and was a man most wise.

Also by Mark Solms

Beyond Evolutionary Psychology (2018, with George Ellis)

The Feeling Brain (2015)

From Neurology to Psychoanalysis (2006, with Lynn Gamwell)

The Brain and the Inner World (2002, with Oliver Turnbull)

Clinical Studies in Neuropsychoanalysis
(2000, with Karen Kaplan-Solms)

The Neuropsychology of Dreams (1997)

A Moment of Transition (1990, with Michael Saling)

Contents

List of Figures

While every effort has been made to contact copyright-holders of illustrations, the author and publishers would be grateful for information about any illustrations where they have been unable to trace them, and would be glad to make amendments in further editions.

Introduction

When I was a child, a peculiar question occurred to me: how do we picture the world as it existed before consciousness evolved? There was such a world, of course, but how do you *picture* it – the world as it was before picturing things became possible?

To give you a sense of what I mean, try to imagine a world in which *a sunrise cannot occur*. The earth has always revolved around the sun, but the sun only rises over the horizon from the viewpoint of an observer. It is an inherently perspectival event. The sunrise will forever be trapped in experience.

This obligatory perspective-taking is what makes it so difficult for us to comprehend consciousness. If we want to do so, we need to elude subjectivity – to look at it from the outside, to see things as they really are as opposed to how they appear to us. But how do we do that? How do we escape our very selves?

As a young man, I naïvely visualised my consciousness as a bubble surrounding me: its contents were the moving pictures and sounds and other phenomena of experience. Beyond the bubble, I assumed there lay an infinite blackness. I imagined this blackness as a symphony of pure quantities, interacting forces and energies and the like: the true reality 'out there' that my consciousness represents in the qualitative forms that it must.

The impossibility of any such imagining – the impossibility of representing reality without representations – illustrates the scale of the task that is tackled in this book. Once again, all these years later, I am trying to peek behind the veil of consciousness, to catch a glimpse of its actual mechanism.

The book you hold in your hands, then, is unavoidably

perspectival. In fact, it is even more perspectival than the paradox I have just described requires it to be. To help you see things from my point of view, I decided to tell a part of my own history. Advances in my scientific ideas about consciousness have often emerged from developments in my personal life and clinical work, and though I believe that my conclusions stand alone, it is much easier to grasp them if you know how I came to them. Some of my discoveries – for example, the brain mechanisms of dreaming – happened largely by serendipity. Some of my professional choices – for example, to take a detour from my neuroscientific career and train as a psychoanalyst – paid off more handsomely than I could reasonably have hoped. In both cases, I will explain how.

But to the extent that my quest to understand consciousness has been successful, my greatest stroke of luck has been the brilliance of my collaborators. In particular, I had the profound good fortune to work with the late Jaak Panksepp, a neuroscientist who, more than any other, understood the origin and power of *feelings*. Pretty much everything that I now believe about the brain was shaped by his insights.

More recently, I have been able to work with Karl Friston, who, among his many excellent qualities, bears the distinction of being the world's most influential living neuroscientist. It was Friston who dug the deepest foundations for the theory I am about to elaborate. He is best known for reducing brain functions (of all kinds) to a basic physical necessity to minimise something called *free energy*. That concept is explained in Chapter 7, but for now, let me just say that the theory that Friston and I have worked out joins with that project – so much so that you may as well call it the free energy theory of consciousness. That's what it is.

The ultimate explanation for sentience is a puzzle so difficult it is nowadays referred to reverentially as 'the hard problem'. Sometimes, once a puzzle is solved, both the question and its answer cease to be interesting. I will leave it to you to judge whether the ideas I set out here shed new light on the hard problem. Either way, I am confident they will help you to see *yourself* in a new light, and to that degree they should remain interesting until such time as they are superseded. After all, in a profound sense, you *are* your consciousness. It therefore seems reasonable to expect a theory of consciousness to explain the fundamentals of why you feel the way you do. It should explain why you are the way you are. Perhaps it should even clarify what you can do about it.

That last topic, admittedly, transcends the intended scope of this book. But it is not beyond the scope of the theory. My account of consciousness unites in a single story the elementary physics of life, the most recent advances in both computational and affective neuroscience and the subtleties of subjective experience that were traditionally explored by psychoanalysis. In other words, the light this theory sheds ought to be light you can use.

It has been my life's work. Decades on, I am still asking myself how the world might have looked before there was anyone around to see it. Now, better educated, I imagine the dawn of life in one of those hydro-thermal vents. The unicellular organisms that came into being there would surely not have been conscious, but their survival prospects would have been affected by their ambient surrounds. It is easy to imagine these simple organisms responding to the biological 'goodness' of the energy of the sun. From there, it is a small step to imagine more complex creatures actively striving for such energy supplies and eventually evolving a capacity to weigh the chances of success by alternative actions.

Consciousness, in my view, arose from the experience of such organisms. Picture the heat of the day and cold of the night from the perspective of those first living beings. The physiological values registering their diurnal experiences were the precursors of the first sunrise.

Many philosophers and scientists still believe that sentience serves no physical purpose. My task in this book is to persuade you of the plausibility of an alternative interpretation. This requires me to convince you that feelings are part of nature, that they are not fundamentally different from other natural phenomena, and that they *do* something within the causal matrix of things. Consciousness, I will demonstrate, is about feeling, and feeling, in turn, is about how well or badly you are doing in life. Consciousness exists to help you do better.

The hard problem of consciousness is said to be the biggest unsolved puzzle of contemporary neuroscience, if not all science. The solution proposed in this book is a radical departure from conventional approaches. Since the cerebral cortex is the seat of intelligence, almost everybody thinks that it is also the seat of consciousness. I disagree; consciousness is far more primitive than that. It arises from a part of the brain that humans share with fishes. This is the 'hidden spring' of the title.

Consciousness should not be confused with intelligence. It is perfectly possible to feel pain without any reflection as to what the pain is about. Likewise, the urge to eat – a feeling of hunger – need not imply any intellectual comprehension of the exigencies of life. Consciousness in its elemental form, namely raw *feeling*, is a surprisingly simple function.

Three other prominent neuroscientists have taken this approach: Jaak Panksepp, Antonio Damasio and Bjorn Merker. Panksepp led the way. He (like Merker) was an animal researcher; Damasio (like me) is not. Many readers will be horrified by the animal research findings I report here, precisely because they show that other animals feel just as we do. All mammals are subject to feelings of pain, fear, panic, sorrow and the like. Ironically, it was Panksepp's research that removed any reasonable doubt on that score. Our only consolation is that his findings made it impossible for such research to continue unabated.

I was drawn to Panksepp, Damasio and Merker because they believed, as I do, that what is lacking in the neuroscience of our time is a clear focus on the embodied nature of *lived experience*. It could be said that what unites us is that we have built, sometimes unwittingly, upon the abandoned foundations that Freud laid for a science of the mind that prioritises feelings over cognition. (Cognition is mostly unconscious.) This is the second radical departure of this book; it returns us to Freud's 'Project' of 1895 – and it attempts to finish the job. But I do not overlook his many mistakes. For one thing, like everyone else, Freud thought that consciousness was a cortical function.

The third and last major departure of this book is that it comes to the view that consciousness is engineerable. It is artificially *producible*. This conclusion, with its profound metaphysical implications, arises from my work with Karl Friston. Unlike Panksepp, Damasio and Merker, Friston is a computational neuroscientist. Therefore, he believes that consciousness is ultimately reducible to the laws of physics (a belief that, surprisingly, was shared by Freud). But even Friston largely equated mental functions with cortical ones before we began our collaboration. This book takes his statistical-mechanical framework deeper, into the most primitive recesses of the brainstem ...

These three departures make the hard problem less hard. This book will explain how.

Mark Solms
Chailey, East Sussex
March 2020

1

The Stuff of Dreams

I was born on the Skeleton Coast of the former German colony of Namibia, where my father administered a small South African-owned company called Consolidated Diamond Mines. The holding company, De Beers, had created a virtual country within a country, known as the *Sperrgebiet* ('prohibited area'). Its sprawling alluvial mines extended from the sand dunes of the Namib Desert down to the Atlantic Ocean floor, several kilometres out to sea.

This was the peculiar landscape that moulded my imagination. As small children, my older brother Lee and I used to play at diamond mining, using toy earth-moving machines, recreating in our garden the impressive engineering feats we witnessed at our father's side when he took us to see the open-cast mines in the desert. (We were, of course, too young to know about the less impressive aspects of his industry.)

One day in 1965, when I was four years old, my parents were yachting at the Cormorant Yacht Club, as they often did, and I was left playing in the clubhouse with Lee, aged six. The early morning mists had burned away. I wandered from the cool interior of the three-storey clubhouse down to the water's edge. Wading there in the heat, I watched tiny shimmering fishes scatter from my feet as Lee and some friends of his clambered onto the roof from the back of the building.

What I remember next are three snapshots. First, the sound of something like a watermelon cracking open. Next, the image of Lee lying on the ground whimpering about a sore leg. Last, my aunt and uncle telling me that they would be looking after my sister and me while our parents travelled to the hospital with Lee. The bit about a sore leg must be a confabulation: the medical records state that my brother lost consciousness upon impact with the concrete paving.

Lee needed specialist care of a kind that our local hospital could not provide. He was flown by helicopter to Groote Schuur Hospital in Cape Town, 800 km away. The neurosurgery department was then housed in an imposing block built in the Cape Dutch style, the very building in which I now work as a neuropsychologist. Lee's skull had fractured and he had suffered an intracranial haemorrhage. When such haematomas expand, they present a life-threatening emergency requiring surgical intervention. My brother was lucky: his resolved over the next few days and he was eventually discharged home.

Apart from the fact that he had to wear a helmet after the accident to protect his fractured skull, Lee looked no different. As a person, however, he was profoundly altered. There is a German word for the feeling this aroused in me, *Unheimlichkeit*, for which there is no adequate English equivalent. Literally, it means 'unhomeliness' but it translates better as 'eeriness' or 'the uncanny'.

The most obvious way in which he was changed was that he lost his developmental milestones. For a time, he even lost reliable bowel control. What I found more disturbing was the fact that he seemed to *think* differently from before. It felt as if Lee was simultaneously there and not there. He seemed to have forgotten many of the games we played. Now our diamond-mining game became simply digging holes. Its imaginative and symbolic aspects no longer spoke to him. He was no longer Lee.

He failed that year at school – his first. The thing I remember most from those early days after the accident was trying to reconcile the dichotomy that my returned brother looked the same but was not the same. I wondered where the earlier version of him had gone.

Over the ensuing years, I fell into a depression. I remember not being able to muster the energy to put on my shoes in the morning, to go to school. This was about three years after the accident. I couldn't find the energy to do these things because I couldn't see the point of them. If our very being depended upon the functioning of our brains, then what would become of me when *my* brain died, with the rest of my body? If Lee's mind was somehow reducible to a bodily organ then, surely, mine was too. This meant that I – my sentient being – would exist only for a relatively short period of time. Then I would disappear.

I have spent my whole scientific career thinking about this problem. I wanted to understand what happened to my brother, and what would in time happen to all of us. I needed to understand what, in biological terms, our existence as experiencing subjects amounted to. In short: to understand consciousness. That is why I became a neuroscientist.

Even in retrospect, I don't believe I could have taken a more direct route to the answers I sought.

The nature of consciousness may be the most difficult topic in science. It matters because you *are* your consciousness, but it is controversial because of two puzzles that have bedevilled thinkers for centuries. The first is the question of how the mind relates to the body – or, for those of a materialist bent (which is almost all neuroscientists), how the brain gives rise to the mind. This is called 'the mind/body problem'. How does the physical brain produce your phenomenal experience? Equally confoundingly, how does the non-physical stuff called consciousness control the physical body?

Philosophers have assigned this problem to what they call 'metaphysics', which is a way of saying they don't think it can be resolved scientifically. Why not? Because science depends upon empirical methods, and 'empirical' implies 'derived from sensory evidence'. The mind is not accessible to sensory observation. It cannot be seen or touched; it is invisible and intangible, a subject, not an object.

The question of what we can know about minds from the outside – how we can even tell when they are present, for that matter – is the second puzzle. It is called 'the problem of other minds'. Simply put: if minds are subjective, then you can only observe your own. How, then, can we know whether other people (or creatures, or machines) have one at all, let alone discern any objective laws governing how minds in general work?

Over the past century, these questions have elicited three major scientific responses. Science relies upon experiments. One thing in our favour is that the experimental method does not aspire to ultimate truths, but rather to what may be described as best guesses. Starting from observations, we offer conjectures as to what might plausibly explain the observed phenomena. In other words, we formulate hypotheses. Then we generate *predictions* from our hypotheses. These take the form: 'if hypothesis X is correct, then Y should happen when I do Z' (where there is a reasonable chance that Y will not happen under some other hypothesis). This is the experiment. If Y does not happen, then X is inferred to be false and is revised in accordance with the new observations. Then the experimental process begins again, until it gives rise to falsifiable predictions that are confirmed. At that point, we hold the hypothesis to be *provisionally* true, until and unless further observations contradict it. In this way, we do not expect to attain certainty in science; we aspire only to less uncertainty.[1]

Starting in the first half of the twentieth century, a school of psychology called 'behaviourism' began systematically to apply the experimental method to the mind. Its starting point was to disregard everything except empirically observable events. The behaviourists threw out all 'mentalistic' talk of beliefs and ideas, feelings and desires, and restricted their field of study to the subject's visible and tangible responses to objective stimuli. They were fanatically uninterested in subjective reports about what was going on inside. They treated the mind as a 'black box', whose inputs and outputs were all that could be known of it.

Why did they take such an extreme stance? Partly, of course, it was an attempt to navigate around the problem of other minds. If they refused to countenance any talk of minds in the first place, it stood to reason that their theories could not be afflicted by the philosophical doubts endemic to psychology. In effect, they excluded the psyche from psychology.

That may seem like a high price to pay. But behaviourism was from the outset a revolutionary doctrine. The behaviourists weren't chasing epistemological purity for its own sake: they were also trying to dethrone the incumbent power in psychology at the time. Freudian psychoanalysis had dominated the science of the mind since the start of the century. By closely examining the curious features of introspective testimonies, Sigmund Freud had sought to develop a model of the mind considered, as it were, from the inside out. The resulting ideas set the agenda for treatment and research for half a century, spawning institutions, accredited experts and a cadre of prominent intellectual champions. Yet in the judgement of the behaviourists, all Freud's theories were just so many cloud castles, erected on the vaporous foundations of subjectivity. Freud had run headlong into the problem of other minds and dragged the rest of psychology after him. It was up to the behaviourists to pull it back again.

Despite the austerity of their programme, they were in fact able to infer causal relations between certain types of mental stimuli and responses. Not only that: they could also manipulate the inputs to elicit predictable changes in the outputs. In doing so, they discovered some of the fundamental laws of learning. For example, when the trigger of an involuntary behaviour is paired repeatedly with an artificial stimulus, then the artificial stimulus will come to trigger the same involuntary response as the innate stimulus. So, if the sight of food is paired repeatedly with the ringing of a bell (in animals that naturally salivate when they see food, as dogs do), then the sound of the bell alone will come to trigger salivation. This is called 'classical conditioning'. Likewise, if a voluntary behaviour is accompanied repeatedly by rewards, that behaviour will increase, and if the same behaviour is accompanied by punishments, it will decrease. So, if a dog that jumps on visitors is hugged, it will jump on them more; if it is smacked, it will jump on them less. This is called 'operant conditioning' – also known as the Law of Effect.

Such discoveries were no small achievement; they showed that the mind is subject to natural laws, like everything else. But there is a lot more to the mind than learning, and even learning is influenced by factors other than external stimuli. Imagine thinking to yourself: 'after I have read this page, I will make myself a cup of tea'. This type of thinking influences your behaviour all the time. Yet the behaviourists did not consider such introspective reports to be acceptable scientific data, because thoughts are not externally observable. In consequence, they could not know what caused you to make your cup of tea.

The great neurologist Jean-Martin Charcot once said: 'theory is good, but it doesn't prevent things from existing'.[2] Since internal mental events clearly do exist and causally influence behaviour, the behaviourist approach was gradually eclipsed in the second half of the twentieth century by another approach. It was called 'cognitive' psychology, which was able to accommodate internal mental processes – in a manner of speaking.

The impetus behind the cognitive revolution was the advent of computers. Behaviourists considered the internal workings of the mind to be an inscrutable 'black box' and focused instead on its inputs and outputs. But computers are not unfathomable. It would have been impossible for us to invent them without thoroughly understanding their inner workings. By treating the mind as though it were a computer, therefore, psychologists felt emboldened to formulate models of the *information processing* that went on within it. Their models were then tested using artificial simulations of mental processes, combined with behavioural experiments.

What is information processing? I will say a lot about it later, but the most interesting thing for our present purposes is that it can be implemented with vastly different kinds of physical equipment. This casts new light on the physical nature of the mind. It suggests that the mind (construed as information processing) is a *function* rather than a structure. On this view, the 'software' functions of the mind are implemented by the 'hardware' structures of the brain, but the same functions can be implemented equally well by other substrates, such as computers. Thus, both brains and computers perform *memory* functions (they encode and store information) and *perceptual* functions (they classify patterns of incoming information by comparing them with stored information) as well as *executive* functions (they execute decisions about what to do in response to such information).

This is the power of what came to be called the 'functionalist' approach, but it is also its weakness. If the same functions can be performed by computers, which presumably are not sentient beings, then are we really justified in reducing the mind to mere information processing? Even your phone has memory, perceptual and executive functions.

The third major scientific response to mind/body metaphysics developed in tandem with cognitive psychology, but by the end of the last century it had grown to overshadow it. I am referring to an approach that is broadly termed 'cognitive neuroscience'. It focuses on the hardware of the mind, and it arose with the development of a plethora of physiological techniques that make it possible for us to observe and measure the dynamics of the living brain directly.

In behaviourist times, neurophysiologists were limited to a single such technique: they could record the brain's electrical activity from the outer surface of the scalp using an electroencephalogram (EEG). Nowadays we have many more tools at our disposal, such as functional magnetic resonance imagery (fMRI) to measure the rates of haemodynamic activity in different parts of the brain while it is performing specific mental tasks, and positron emission tomography (PET), with which we can measure differential metabolic activity for single neurotransmitter systems. This enables us to identify precisely which brain processes generate our different mental states. We can also visualise the detailed functional-anatomical connectivity between those different brain regions using diffusion tensor tractography. And by using optogenetics we can see and activate the circuits of neurons comprising individual memory traces as they light up during cognitive tasks.

These techniques render the inner workings of the organ of the mind plainly visible – thereby realising the wildest empiricist dreams of the behaviourists without limiting the scope of psychology to stimuli and responses.

The state of neuropsychology in the 1980s when I entered the field explains why behaviourists made such a seamless transition from learning theory to cognitive neuroscience. The neuropsychology of that time might as well have been called neurobehaviourism. The more I was taught about functions like short-term memory, which was said to provide a 'buffer' for holding memories in consciousness, the more I realised that my lecturers were talking about something other than what I had signed up for. They were teaching us about the functional tools used by the mind, rather than the mind itself. I was dismayed.

The neurologist Oliver Sacks, in his book *A Leg to Stand On* (1984), aptly described the situation I found myself in:

> Neuropsychology, like classical neurology, aims to be entirely objective, and its great power, its advances, come from just this. But a living creature, and especially a human being, is

first and last active – a subject, not an object. It is precisely the subject, the living 'I', which is being excluded. Neuropsychology is admirable, but it excludes the psyche – it excludes the experiencing, active, living 'I'.[3]

That line 'Neuropsychology is admirable, but it excludes the psyche' captured my disappointment perfectly. Upon reading it, I entered into a correspondence with Oliver Sacks that continued until his death in 2015. What drew me to him was the fact that he took so seriously the subjective reports of his patients. This was evident already in his 1970 book *Migraine*, and even more so in his extraordinary *Awakenings* (1973). The second book recorded in exquisite detail the clinical journeys of a group of chronic 'akinetic-mute' patients with encephalitis lethargica. This disease was also known as 'sleeping sickness', although the patients were not literally asleep, rather they showed no spontaneous initiative or drive. Sacks 'awakened' them by giving them levodopa, a drug that increases the availability of dopamine. Following the return of active agency, however, they rapidly became excessively driven, manic and eventually psychotic. Shortly after I read *A Leg to Stand On*, which described Sacks's own subjective experience of a nervous-system injury, he published *The Man Who Mistook His Wife for a Hat* (1985) – a series of case studies that provided enlightening insights into neuropsychological disorders from the perspective of *being* a neurological patient. This brought Sacks lasting fame.

These books were quite unlike my neuropsychological textbooks, which dissected mental functions as we would the functions of any bodily organ. For example, I learnt that language was produced by Broca's area in the left frontal lobe, that speech comprehension took place in Wernicke's area, a few centimetres further back, in the temporal lobe, and that the ability to repeat what is said to you was mediated by the arcuate fasciculus, a fibre tract that connects these two regions. Likewise, I learnt that memories were encoded by the hippocampus, stored in the neocortex and retrieved by frontal-limbic mechanisms.

Was the brain really no different from the stomach and lungs? The obvious thing that set it apart was the fact that there is 'something it is like' to *be* a brain. This did not apply to any other part of the body. The sensations that we locate in other bodily organs are not felt by the organs themselves; nerve impulses arising from them are felt only when they reach the brain. Surely this highly distinctive property of brain tissue – the capacity to sense, feel and think things – existed for a reason. This property appeared to do something. And if it did – if subjective experience had causal effects upon behaviour, as it seems to when we spontaneously decide to make a cup of tea – then we would be led badly astray if we omitted it from our scientific accounts. Yet that is precisely what was happening in the 1980s. At no point did my lecturers say anything about what it is like to comprehend speech or retrieve a memory, let alone why it feels like anything at all.

Those who did take the subjective perspective into account were not taken seriously by proper neuroscientists. I am not sure how many people know that Sacks's publications were widely derided by his colleagues. One commentator went so far as to call him 'The man who mistook his patients for a literary career'. This caused him a good deal of distress. How can you describe the inner life of human beings without telling their stories? As Freud had lamented a century before in relation to his own clinical reports:

> It still strikes me as strange that the case histories I write should read like short stories and that, as one might say, they lack the serious stamp of science. I must console myself with the reflection that the nature of the subject is evidently responsible for this, rather than any preference of my own.[4]

Sacks was delighted when I sent him this quotation.[5] For my own part, when I first read these lines, I realised that I was not alone in having entered neuropsychology with the hope that it would enable me to learn how the brain generates subjectivity. One is quickly disabused of this notion. You are warned not to pursue such intractable questions – they are 'bad for your career'. And so, most students of neuroscience gradually forget why they entered the field, and come to identify with the dogma of cognitivism, which approaches the brain as though it were no different from a mobile phone.

The one aspect of consciousness that *was* a respectable scientific topic in the 1980s was the brain mechanism of wakefulness versus sleep. In other words, the 'level' of consciousness was a respectable topic but not its 'contents'. So, I decided to focus my doctoral research on an aspect of sleep. In particular, I chose to study the subjective aspect of sleep, namely the brain mechanisms of dreaming. Dreaming, after all, is nothing but a paradoxical intrusion of consciousness ('wakefulness') into sleep. Amazingly, there was a huge gap in the literature on this topic: nobody had systematically described how damage to different parts of the brain affected dreaming. So, this is what I set out to do.

What makes dreaming tricky to study is precisely its subjective nature. Mental phenomena in general can be witnessed only introspectively by a single observer and then reported to others indirectly, through words. But dreams are even more problematic: they can be reported only retrospectively, once the dream is over and the dreamer has woken up. Everyone knows how unreliable our memory for dreams is. What kind of 'data' are those?[6] Which is why, from the middle of the twentieth century onwards, dreams were a significant front in the transition from behaviourism to what would later become cognitive neuroscience.

The electroencephalogram was first applied to the study of sleep in the early 1950s by two neurophysiologists, Eugene Aserinsky and Nathaniel Kleitman. They hypothesised that the level of brain activity would decrease as we fall asleep and increase when we wake up, and therefore predicted that the amplitude of our brainwaves (which is one of the things that electroencephalography measures) would increase and their frequency (the other thing it measures) would decrease as we fall asleep; and that the opposite would happen when we wake up (see Figure 10 on p. 127).

When the brain descends into what is now called 'slow wave' sleep, we see exactly what Aserinsky and Kleitman predicted. Their hypothesis was confirmed. The surprise is what happens next: within about ninety minutes of drifting off (and roughly every ninety minutes thereafter, in regular cycles) the brainwaves speed up again, almost reaching waking levels, even though the person from whom the recordings are being obtained remains asleep.[7] Aserinsky and Kleitman named these curious states of brain activation 'paradoxical sleep' – the paradox being that the brain is physiologically aroused despite being fast asleep.

Various other things happen in this peculiar state. The eyes move rapidly (which is why paradoxical sleep was later renamed 'rapid eye movement' or REM sleep), yet the body below the neck is temporarily paralysed. There are dramatic autonomic changes, too, such as reduced control of core body temperature and engorgement of the genitals leading to visible erections in men. How science managed not to notice all this until 1953 is mind-boggling.

On the basis of these observations, Aserinsky and Kleitman formulated a further, not-unreasonable hypothesis: that REM sleep is the physiological basis of the psychological state called dreaming. Accordingly, they predicted that awakenings from REM sleep would elicit dream reports while awakenings from slow-wave (non-REM) sleep would not. Together with the unfortunately named William Dement, they tested this prediction and confirmed it: whereas approximately 80 per cent of awakenings from REM sleep produced dream reports, fewer than 10 per cent of awakenings from non-REM sleep did so. From that moment onward, REM sleep was considered to be synonymous with dreaming.[8] Excellent news! The field no longer had to bother with dreaming, because now we had an objective marker of it, which enabled neuroscientists to do proper science without having to contend with the methodological complications introduced by retrospective, single-witness, verbal reports of fleeting subjective experiences.

There was another reason to be grateful for getting rid of dreams. This was the embarrassing role that they had played in the establishment of psychoanalysis. Unlike the mainstream responses to mind/body metaphysics that characterised mental science in the second half of the twentieth century, psychoanalysts had no qualms about treating introspective reports as data. In fact, reports elicited by 'free association' (unstructured sampling of the stream of consciousness) were the primary data of psychoanalytic research. Using this method, Sigmund Freud came to the conclusion that, despite the nonsensical appearance of 'manifest' dream experiences, their 'latent' content (the underlying story, which he inferred from the dreamer's free associations) revealed a coherent psychological function. This function was *wish-fulfilment*.

According to Freud, dreaming is what happens when the biological needs that generate waking behaviour are released from inhibition during sleep. Dreams are attempts to meet those needs, which continue to make demands upon us even when we sleep. However, dreams do so in a hallucinatory fashion, and thereby enable us to stay asleep (rather than wake up in order to really satisfy our drives). Since hallucinations are a core feature of mental illness, Freud in his seminal book *The Interpretation of Dreams* (1900) used this theory to paint a broad-brushstroke model of how the mind as a whole works, in health and disease.

As Freud put it: 'psychoanalysis is founded upon the analysis of dreams'.[9] But dreams, as we have seen, are incredibly difficult things to study empirically, and so the behaviourists ruled them out of science. What was more, the theoretical edifice that Freud built upon dreams was no better than its foundations. The great philosopher of science Karl Popper declared psychoanalytic theory 'pseudoscientific', because it did not give rise to experimentally falsifiable predictions.[10] How do you falsify the hypothesis that dreams express the latent desires that Freud inferred? If the desires do not have to appear in the manifest (reported) dream, then any dream can be 'interpreted' to suit the requirements of the theory. Not surprisingly, therefore, when the discovery of REM sleep made it possible for neuroscientists to shift from the ephemeral stuff of dream reports to their concrete physiological correlates, the dreams themselves were dropped like slippery fish.

The discovery of REM sleep in the 1950s triggered a race to identify its neurological basis, since the function of REM sleep could reveal the *objective* mechanism of dreams, whose elucidation would place the psychiatry of the time on a more respectable scientific footing. (This research was made easier by virtue of the fact that REM sleep occurs in all mammals.) The race was won by Michel Jouvet, in 1965. In a series of surgical experiments on cats, he demonstrated that REM sleep was generated not by the forebrain (which includes the cortex, the upper part of the brain that is so impressively large in humans and partly for that reason is considered the organ of the mind) but rather by the brainstem, a supposedly much humbler structure of exceedingly ancient evolutionary origin.[11] Jouvet came to this conclusion by observing that progressive slices through the brain, starting at the top and working downwards, only produced loss of REM sleep once the cutting had reached the level of a 'lowly' brainstem structure known as the pons (see Figure 1).[12]

It fell to Jouvet's student Allan Hobson to wrap up the details. Hobson identified precisely which assemblies of pontine neurons generated REM sleep and therefore dreams. It became apparent by the mid-1970s that the whole sleep/waking cycle – including all the phenomena of REM sleep enumerated above, as well as those of the different stages of non-REM sleep – were orchestrated by a small number of brainstem nuclei interacting with each other.[13]

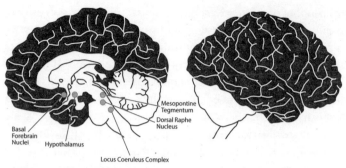

Figure 1 The image on the left is a medial view of the brain (cut through the middle) and the one on the right is a lateral view (seen from the side). The figure shows the cortex (black) and brainstem (white). Only those brainstem nuclei considered important for REM sleep control are indicated, namely the mesopontine tegmentum, dorsal raphe nucleus and locus coeruleus complex. Also shown are the location of the basal forebrain nuclei (underneath the cortex) and the hypothalamus, the relevance of which becomes apparent later.

Those controlling REM sleep resembled a simple on/off switch. The neurons that switch REM *on* are found in the mesopontine tegmentum (see Figure 1). They release a neurochemical called acetylcholine throughout the forebrain. Acetylcholine causes arousal: it increases the 'level' of consciousness (for example, it is boosted by nicotine, which thereby helps you concentrate). The brainstem neurons that switch REM sleep *off* are located deeper within the pons, in the dorsal raphe and locus coeruleus complex (again, see Figure 1). They release serotonin and noradrenaline respectively. Like acetylcholine, these neurochemicals modulate different aspects of the level of consciousness.

Combining these findings with the fact that REM sleep switches on and off automatically, roughly every ninety minutes, like clockwork, Hobson wasted no time in drawing the inevitable conclusion: 'The primary motivating force for dreaming is not psychological but physiological since the time of occurrence and duration of dreaming sleep are quite constant, suggesting a pre-programmed, neurally determined genesis.'[14]

Because REM sleep arises from the cholinergic brainstem, an ancient and lowly part of the brain far from the majestic cortex where all the action of human psychology presumably takes place, he added that dreaming could not possibly be motivated by wishes; it was 'motivationally neutral'.[15] Therefore, according to Hobson, Freud's view that dreams were driven by latent desires must be completely wrong. The meaning that Freud saw in dreams was no more intrinsic to them than it is to inkblots. It was projected onto them; it was not in the dream itself. From the scientific point of view, dream interpretation was no better than reading tea leaves.

Because the whole of psychoanalysis was grounded on the method that Freud used to study dreams, the entire body of theory that he derived in this way could be dismissed. Following Hobson's demolition job on the idea that dreams might mean anything at all, psychiatry could at last turn away from its historical reliance on introspective reports and base itself instead upon objective neuroscientific (especially neurochemical) methods of research and treatment. In consequence, whereas in the 1950s it was almost impossible to become a tenured professor of psychiatry at a leading American university unless you were a psychoanalyst, today the opposite is true: it is almost impossible to become a professor of psychiatry if you *are* a psychoanalyst.

None of this particularly struck me at the time. The question at the heart of my doctoral research seemed fairly straightforward, and not at all implicated in the battles over the legacies of Freudianism and behaviourism. All I wanted to know was this: how did damage to different parts of the forebrain and its cortex affect the actual experience of dreaming? After all, if the forebrain was where the action was, psychologically speaking, surely it must do *something* in dreams.

The department of neurosurgery at the University of the Witwatersrand had wards in two teaching hospitals – Baragwanath Hospital and Johannesburg General Hospital. Baragwanath was a sprawling ex-military hospital, set in the 'non-European' township of Soweto. Bearing in mind that this was during the height of apartheid in South Africa, it was a sea of human misery. The Johannesburg General Hospital, by contrast, which was reserved for 'Europeans', was a state-of-the-art academic hospital; a monument to racial inequality. The neurosurgery department also had beds in the Brain and Spine Rehabilitation Unit at Edenvale General Hospital, which was in an old colonial building set in Johannesburg's suburbia. Starting in 1985 I worked across all three sites, examining hundreds of patients per year. I included 361 of them in my doctoral research, which extended over the next five years.

After learning how to use electroencephalographic and related technology and to recognise the characteristic brainwaves associated with the different stages of sleep, I was able to wake people up during REM, when they were most likely to be dreaming. I also asked neurological patients at the bedside about changes in their dreams, and then followed them up over days, weeks and months. This is how I proceeded to investigate whether the content of dreams was systematically affected by localised damage to different parts of the brain. Despite the dubious reputation of dream reports, I assumed that if patients with damage to the same brain area claimed the same change in dream content, there was every reason to believe them. This method is called 'clinico-anatomical correlation': by probing the psychological capacities of patients clinically, you observe how a mental function has been altered by damage to a part of the brain; then you correlate that alteration with the site of the damage, in this way discovering clues about the function of the damaged brain structure, which leads to testable hypotheses. The method had been systematically applied decades before to all the major cognitive functions, such as perception, memory and language, but it had not yet been applied to dreaming.

At first, I was a little uneasy about talking to such seriously ill people about their dreams. Many of them were facing, or had just undergone, life-threatening brain surgery, and in the circumstances I feared they might consider my questions frivolous. But my patients were surprisingly willing to describe the changes in their mental life that neurological diseases had brought about.

By the time I began my research, several case reports had been published in which the same effect observed in experimental animals was shown to occur in human beings: namely that REM sleep was obliterated by damage to the mesopontine tegmentum (see Figure 1). But, astonishingly, nobody had bothered to enquire about changes in these patients' *dreams*. This is as clear an example as one can get of the prejudice against subjective data in neuroscience.[16]

In my research, I expected to find the obvious: that patients with damage to the visual cortex would experience non-visual dreams; that patients with damage to language cortex would experience non-verbal dreams; that patients with damage to somatosensory and motor cortex would experience hemiplegic dreams; and so on. These are the ABCs of brain/behaviour correlation. This was the gap I wanted to fill; and, happily, I did.[17]

To my amazement, however, alongside all the obvious things I observed, I found also that patients with damage to the part of the brain that generates REM sleep *still experienced dreams*. Moreover, patients in whom dreaming *was* abolished had damage to a completely different part of the brain. Dreaming and REM sleep were therefore what we call 'doubly dissociable' phenomena.[18] They were correlated with each other (i.e. they usually happened at the same time) but they were not the same thing.[19]

For a period of almost fifty years, in the whole field of sleep science, brain researchers had been confusing correlation with identity. As soon as they had established that dreaming accompanied REM sleep, they leapt to the conclusion that they were one and the same – then jettisoned the troublesome subjective side of the correlation. Thereafter, with very few exceptions, they studied REM sleep alone, mainly in experimental animals, which cannot provide introspective reports. The error came to light only when I began to take neuroscientific interest in the *experience* of dreams in neurological patients.

When, in the early 1990s, I first reported that dreaming was obliterated by damage in a different part of the brain from the part that generates REM sleep, I took pains to stress that the critical area was not in the brainstem.[20] This was because I wanted to emphasise the mental nature of dreaming, and we all knew that mental functions reside in the cortex.

In fact, I found two areas of damage that caused loss of dreaming with preservation of REM sleep. The first was in the cortex, in the inferior parietal lobule (see Figure 2). That finding was not surprising, as the parietal lobe is important for short-term memory. If a patient cannot hold the contents of their memory in the buffer of consciousness, how can they experience a dream? Far more interesting was the second brain area, namely the white matter of the ventromesial quadrant of the frontal lobes, which connects the frontal cortex to various subcortical structures. This finding was totally unexpected; nothing about the functions of this part of the brain is obviously connected with the manifest experience of dreaming, and yet it must contribute something crucial to the process, because damage there reliably caused a total cessation of dreaming.

I say 'reliably', even though I reported only nine instances of loss of dreaming among my frontal-lobe patients (and forty-four cases of the parietal type of damage). Such injuries are extremely rare in ordinary clinical practice. Nevertheless, the correlation is reliable. In the first half of the twentieth century, the ventromesial frontal white matter was targeted surgically in thousands of cases by a technique known as modified prefrontal leucotomy.

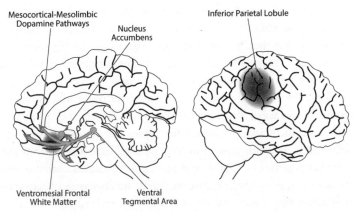

Mesocortical-Mesolimbic
Dopamine Pathways

Nucleus
Accumbens

Inferior Parietal Lobule

Ventromesial Frontal
White Matter

Ventral
Tegmental Area

Figure 2 The two areas of damage which lead to cessation of dreaming are shaded in this figure, namely the ventromesial white matter of the frontal lobe (left) and the inferior parietal lobule of the cortex (right). Also shown is the ventral tegmental area of the brainstem and the major fibre pathways arising from it, namely the mesocortical-mesolimbic dopamine pathways. Please note: the lesion site in the ventromesial frontal lobe involves these subcortical pathways, which course beneath the cortex, not within it. A major destination of these pathways is the nucleus accumbens, which is also shown.

The psychiatrists of those gung-ho days discovered that some serious mental illnesses could be ameliorated by complete surgical destruction of the prefrontal lobes (technically known as frontal lobotomy); but they also noticed that this radical procedure had many 'side effects', as they euphemistically called them. They therefore reduced the extent of the lesion, trying to identify what was the smallest part of the frontal lobes that could be disconnected from the rest of the brain to still obtain the desired results. The modified procedure of Walter Freeman and James Watts was the answer. It required insertion of a tiny whirring blade through the eye sockets, which cut the white matter in the ventromesial quadrant of the frontal lobes (prefrontal leucotomy), in the precise location of the damage in my nine patients.

I therefore went back to the old psychosurgical literature to see if it confirmed what I observed in my cases.[21] I had reason to be hopeful that the doctors who examined the classical leucotomy patients would have asked them about dreams after their operations; after all, dreams were still taken seriously by psychiatrists back then. It turned out I was right. What they found is that prefrontal leucotomy had three main psychological effects. First, it reduced positive psychotic symptoms (hallucinations and delusions). Second, it reduced motivation. Third, it caused loss of dreaming. In fact, one of the early psychosurgical investigators went so far as to suggest that preservation of dreaming after the operation was a bad prognostic sign.[22]

This last point helped me to conjecture which of the many neural circuits in the ventromesial quadrant of the frontal lobes was most likely to be responsible for the loss of dreaming. It also provided a first hint as to why we should find our culprit in this unexpected region of the brain. What are dreams if not hallucinations and delusions? That is why it would be a bad prognostic sign if they persisted after leucotomy.

As it happens, the neurosurgical treatment of hallucinations and delusions was not abandoned for ethical reasons; it fell out of favour when it became apparent that equivalent therapeutic results could be obtained with less morbidity and mortality by using some drugs which first became widely available in the 1950s, namely 'major tranquillisers'. What these drugs did, and modern 'antipsychotics' still do, was block the neurochemical dopamine at the terminals of a brain circuit known as the mesocortical-mesolimbic dopamine system (see Figure 2). Since this circuit is cut through by prefrontal leucotomy, as it was in my nine patients with naturally occurring damage, I hypothesised that this might be the system that generates dreams.

Further experiments confirmed my hypothesis. It had already been established that pharmacological stimulation of this circuit increased the frequency, length and intensity of dreams, without commensurate effects on REM sleep.[23] The drug in question was levodopa, the very same drug that Oliver Sacks had used to 'awaken' his post-encephalitic patients. Neurologists using dopamine stimulants for the treatment of Parkinson's disease have long known that they must be careful not to push their patients into psychosis, like Sacks did; and the onset of unusually vivid dreams is often the first sign of this side effect.[24] The crucial subsequent observations were that the neurons that constitute this circuit (the cell bodies of which are located in the ventral tegmental area) fire at maximum rates during dreaming sleep,[25] and at the same time deliver dopamine in maximum quantities to their targets in the nucleus accumbens (see Figure 2).[26] It is therefore now widely accepted that dreaming can occur independently of REM sleep and that the mesocortical-mesolimbic dopamine circuit is indeed the major driver of dreaming.[27]

Damage to cholinergic pathways in the ventromesial quadrant of the frontal lobes (which arise from the basal forebrain nuclei, see Figure 1) produces the opposite effect to what happens when dopaminergic pathways are damaged, namely *more* dreaming rather than less. Hobson had claimed that acetylcholine was the motivationally neutral generator of dreams, but the same thing occurs if you block acetylcholine pharmacologically as happens when its pathways are damaged. Anticholinergic drugs – acetylcholine blockers – are now widely known to cause *excessive* dreaming.[28] In other words, blocking of the neural system that Hobson claimed was responsible for dreaming has the opposite effect to what his theory predicted.

It rapidly became clear that neuroscience owed Freud an apology. If there is one part of the brain that might be considered responsible for 'wishes', it is the mesocortical-mesolimbic dopamine circuit. It is anything but motivationally neutral. Edmund Rolls (and many others) calls this circuit the brain's 'reward' system.[29] Kent Berridge calls it the 'wanting' system. Jaak Panksepp calls it the SEEKING system – and foregrounds its role in the function of *foraging*.[30] This is the brain circuit responsible for 'the most energised exploratory and search behaviours an animal is capable of exhibiting'.[31] It is also the circuit that drives dreaming.[32]

Hobson was not amused. He invited me to present my findings to his research group in the department of neurophysiology at Harvard. Initially he accepted them, and published a favourable review of the book I wrote on the topic in 1997, noting that my clinico-anatomical findings were confirmed down to the last detail by Allen Braun's neuroimaging studies (see Figure 3, p. 37).[33] Then he realised that these developments might vindicate a broadly Freudian outlook on dreams, at which point he wrote to me saying that he was willing to endorse my findings publicly only on the condition that I did not claim they supported Freud. So much for the supposed objectivity of neuropsychology.

Yet there was one other very surprising aspect to my discovery. When I first stumbled upon it, I did not pay much attention to the fact that the neurons which drive this circuit are located in the *brainstem* (like those of the circuits that generate REM sleep). As I said, I wanted to emphasise the mental nature of dreaming. My oversight had to be politely pointed out to me by Allen Braun, the neuroimager just mentioned. In the context of the scientific disagreement between myself and Hobson as to which brain circuits drive the dream process (dopaminergic or cholinergic), Braun wrote:

> The curious thing is that, after making a case that forebrain structures must play a critical role in the dream system, Solms ends up by suggesting that it is the dopaminergic afferents to these regions that [generate dreams] – *thereby placing the dream instigator back in the brainstem.*[34]

Braun concluded: 'It sounds to me like these gentlemen are approaching common ground.'[35] In the 1990s, in common with the rest of neuropsychology, I thought the cortex was where all the psychological action was, so I focused on the fact that the white matter tracts that interested me were in the frontal lobes, which is where the damage in my nine cases was located. But all the core nuclei of the brainstem send long axons upwards into the fore-brain (see Figure 2). The cell bodies of these neurons are located in the brainstem, although their output fibres (the axons) termin-ate in the cortex. This underpins the main *arousal* function of these brainstem nuclei, known collectively as the reticular activat-ing system. It was these activating pathways that were damaged in my nine patients, and in the hundreds of documented non-dreaming leucotomy patients who preceded them.

From 1999 onwards, partly prompted by Braun's comments about the implications of my discovery, I directed my attention to the other arousal systems of the brainstem. The most interest-ing work in this area was being done by Jaak Panksepp, whose encyclopaedic book *Affective Neuroscience* (1998) laid out in exquisite detail a vast array of evidence for his view that these supposedly mindless systems, responsible for regulating only the 'level' of consciousness, generated a 'content' of their own.

This would turn out to be highly significant.

2

Before and After Freud

In 1987 I made another decision that put me at odds with the rest of my field. I decided to train as a psychoanalyst.[1] My emerging dream-research findings had convinced me that subjective reports had a vital role to play in neuropsychology, and that the field's opposition to Freud had led it into error in more ways than one. But my research findings weren't the deciding factor.

What made up my mind was a seminar that I attended at the University of the Witwatersrand, in the mid-1980s, led by a professor of comparative literature named Jean-Pierre de la Porte. The seminar concerned *The Interpretation of Dreams*, which I was curious about in light of my doctoral research. Like everybody else in those days, I was sceptical about Freud. I had learnt since my undergraduate years that psychoanalysis was 'pseudoscience'. Nobody in the hard sciences took Freud seriously any more, which is presumably why the seminar took place in a humanities department. The reason I attended was Freud had been willing to talk about the *content* of dreams, the topic of my research.

De la Porte explained that one could not understand the theoretical conclusions Freud reached without first digesting an earlier manuscript of his, written in 1895 but published only in the 1950s, after his death. This manuscript was titled 'Project for a scientific psychology'.[2] In it, Freud attempted to place his early insights about the mind on a neuroscientific footing.

In doing so, he was following in the footsteps of his great teacher, the physiologist Ernst von Brücke, a founding member of the Berlin Physical Society. The mission of this society was formulated as follows by Emil du Bois-Reymond in 1842:

> Brücke and I pledged a solemn oath to put into effect this truth: 'No other forces than the common physical and chemical ones are active within the organism. In those cases which cannot currently be explained by these forces one has either to find the specific way or form of their action by means of the physical-mathematical method or to assume new forces equal in dignity to the chemical-physical forces inherent in matter, reducible to the forces of attraction and repulsion.'[3]

Their beloved teacher, Johannes Müller, had asked how and why organic life differs from inorganic matter. He concluded that 'living organisms are fundamentally different from non-living entities because they contain some non-physical element or are governed by different principles than are inanimate things'.[4] In short, according to Müller, living organisms possess a 'vital energy' or 'life force', which physiological laws cannot explain. He held the view that living creatures cannot be reduced to their component physiological mechanisms because they are indivisible wholes with *aims* and *purposes*, which Müller attributed to the fact that they possess a soul. Considering that the German word *Seele* can be translated as either 'soul' or 'mind',[5] the disagreement between Müller and his students bears a striking resemblance to the debate raging in our own time between philosophers like Thomas Nagel and Daniel Dennett as to whether *consciousness* can be reduced to physical laws (Nagel claims it cannot, Dennett claims it can).

The surprise for me, upon attending De la Porte's seminar, was to learn that Freud – the pioneering investigator of human subjectivity – had aligned himself not with the vitalism of Müller but rather with the physicalism of Brücke. As he wrote in the opening lines of his 1895 'Project': 'The intention is to furnish a psychology that shall be a natural science: that is, to represent psychical processes as quantitatively determinate states of specifiable material particles.'[6]

I hadn't realised that Freud was a neuroscientist. Now I learnt that he had only reluctantly abandoned neurological methods of enquiry when it became clear to him, somewhere between 1895 and 1900, that the methods then available were not up to the task of revealing the physiological basis of mind.

Freud's change of heart brought ample compensation, though. It forced him to look more closely at psychological phenomena in their own right, and to elucidate the functional mechanisms that underpinned them. This gave rise to the psychological mode of investigation that he went on to call 'psychoanalysis'. Its fundamental assumption was that manifest (nowadays called 'explicit' or 'declarative') subjective phenomena have latent (nowadays called 'implicit' or 'non-declarative') causes. That is, Freud argued that the erratic train of our conscious thoughts can be explained only if we assume implicit intervening links of which we are unaware. This gave rise to the notion of latent mental functions and, in turn, to Freud's famous conjecture of 'unconscious' intentionality.

Since no methods were available at the turn of the nineteenth century to investigate the physiology of unconscious mental events, their mechanisms could be inferred only from clinical observation. What Freud learnt in this way gave rise to his second fundamental claim. He observed that patients adopted a far-from-indifferent attitude to their inferred unconscious intentions; it appeared to be more a matter of being *unwilling* rather than unable to become aware of them. He called this tendency variously 'resistance', 'censorship', 'defence' and 'repression', and observed that it prevents emotional distress. This in turn revealed the pivotal role that *feelings* play in mental life, how they underpin all sorts of self-serving biases. These findings (obvious today) showed Freud that some of the major motivating forces in mental life are entirely subjective but also unconscious. Systematic investigation of those forces led him to his third fundamental claim. He concluded that what ultimately underpinned feelings were bodily needs; that human mental life, no less than that of animals, was *driven* by the biological imperatives to survive and reproduce. These imperatives, for Freud, provided the link between the feeling mind and the physical body.

Freud took a remarkably subtle approach to the mind/body relationship. He realised that the psychological phenomena he studied were not straightforwardly reducible to physiological ones. As early as 1891 he argued that it was not possible to attribute psychological symptoms to neurophysiological processes without first reducing the relevant psychological and physiological phenomena (both sides of the equation) to their respective underlying *functions*. As noted earlier with reference to information processing, functions can be performed on various substrates.[7] It was only upon the common ground of function, Freud argued, that psychology and physiology can be reconciled. His goal was to explain psychological phenomena by means of 'metapsychological' functional laws (the term means 'beyond psychology').[8] Trying to skip over this functional level of analysis, jumping directly from psychology to physiology, is nowadays called the localisationist fallacy.[9]

Clearly, for Freud if not his followers, psychoanalysis was meant to be an interim step. Although his quest from the first had been to discern the laws underpinning our rich inner life of subjective experience, nevertheless mental life remained a *biological* problem for him.[10] As he wrote in 1914: 'all our provisional ideas in psychology will presumably someday be placed on an organic foundation'.[11] He therefore enthusiastically anticipated the day when psychoanalysis would once again join up with neuroscience:

> Biology is truly a land of unlimited possibilities. We may expect it to give us the most surprising information, and we cannot guess what answers it will return in a few dozen years

[...] They may be of a kind which will blow away the whole of our artificial structure of hypothesis.[12]

This was not the wildly speculative Freud that I had learnt about as an undergraduate student. The 'Project' was a revelation to me, as it had been to Freud himself. He wrote to his friend Wilhelm Fliess at the time:

In the course of a busy night [...] the barriers were suddenly raised, the veils fell away, and it was possible to see through from the details of the neuroses to the determinants of consciousness. Everything seemed to fit together, the gears were in mesh, the thing gave one the impression that it was really a machine and would soon run of itself.[13]

But the euphoria lasted only a short time. A month later he wrote: 'I can no longer understand the state of mind in which I hatched the "Psychology"; I cannot make out how I came to inflict it on you.'[14] Devoid of appropriate neuroscientific methods, Freud relied upon 'imaginings, transpositions and guesses' to translate his clinical inferences first into functional and then into physiological and anatomical terms.[15] After a final attempt at a revision (contained in a long letter that he sent to Fliess on 1 January 1896), the 'Project' disappeared from view until its re-emergence some fifty years later. But the ideas contained in it (its 'hidden ghost', according to Freud's translator James Strachey), haunted the whole of his psychoanalytic theorising – awaiting future scientific progress.[16]

The 'Project' contained two ideas that stand out in light of contemporary findings. Firstly, the forebrain is a 'sympathetic ganglion' monitoring and regulating the needs of the body. Secondly, these needs are the driving force of mental life, 'the mainspring of the psychical mechanism'.[17] Lacking any neurobiological understanding of how such bodily needs are regulated in the brain – let alone how they could be explained 'by means of the physical-mathematical method' – Freud had no choice but to 'assume new forces equal in dignity to the chemical-physical forces inherent in matter' if he was going to remain true to the ideals of the Berlin Physical Society. These were what he called 'metapsychological' forces, the forces that *lie behind* psychological phenomena. He clarified that he wanted to 'transform metaphysics into meta-psychology'.[18] In other words, he wanted to replace philosophy with science – a science of subjectivity. He asked us not to judge his speculative inferences concerning latent mental processes too harshly:

> This is merely due to our being obliged to operate with the scientific terms, that is to say with the figurative language, peculiar to psychology (or, more precisely, to depth psychology). We could not otherwise describe the processes in question at all, and indeed we could not even become aware of them. The deficiencies in our description would probably vanish if we were already in a position to replace the psychological terms by physiological and chemical ones.[19]

Foremost among the new forces that Freud was obliged to infer was the concept of 'drive', which he defined as 'the psychical representative of the stimuli originating from within the organism and reaching the mind, as *a measure of the demand made upon the mind for work* in consequence of its connection with the body'.[20]

Freud's notion of 'drive' – which he considered to be the source of all 'psychical energy' – was not unlike Müller's 'vital energy', but it was rooted in bodily needs. Freud described the causal mechanisms by which drives become intentional cognition as an 'economics of nerve-force'.[21] Still, he freely admitted that he was 'totally unable to form a conception' of how bodily needs become a mental energy.[22]

When I read these words, about a century later, I realised that the time had now arrived for us to 'replace the psychological terms by physiological and chemical ones'. For example, the driving force behind dreams, which was 'latent' in the subjective reports of Freud's patients and the existence of which was therefore considered unfalsifiable, was clearly 'manifest' in the objective evidence obtained by modern in vivo physiological methods that were not available in Freud's day. Consider for example the images in Figure 3, derived by positron emission tomography;[23] they clearly show that the 'wishful' SEEKING circuit lights up like a Christmas tree during dreaming sleep, while the inhibitory prefrontal lobes are essentially switched off. On the basis of these findings, when Hobson and I were invited to debate the scientific credibility of Freudian dream theory at the 2006 Science of Consciousness conference, our assembled colleagues voted 2:1 in favour of the theory being viable once more.[24]

The subjective 'I' was never excluded from psychoanalysis, for all its faults. It had pride of place, no matter how embarrassing that was for the rest of science. Many scientific colleagues advised me not to associate what I was doing with psychoanalysis, given the historical baggage the word carried. They said it was like an astronomer associating himself with astrology. But I considered it intellectually dishonest to not give Freud his due. No matter how incompletely he achieved his goals, they were clearly the correct *goals* for a science of the mind. So, I called my approach 'neuropsychoanalysis'. I have said that the neuropsychology I was taught might as well have been called neurobehaviourism, such was its attitude to subjectivity. I wanted to be clear that the neuropsychology I was developing pivoted on lived experience. In that spirit, after writing a programmatic paper on the relationship between psychoanalysis and neuroscience, I set to work.[25]

I moved to London in 1989, to undergo psychoanalytic training. To continue my research and clinical work, I simultaneously took

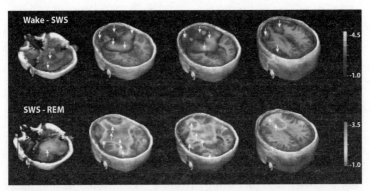

Figure 3 The horizontal rows show progressively higher slices through the brain (from left to right). The top row shows the difference between the awake and sleeping brain, with the shaded area depicting *decreased* cortical activation with sleep onset; the bottom row shows the difference between REM and non-REM (slow-wave) sleep, with the highlighted area depicting *increased* subcortical activation with REM onset. The area of greatest activation is where the SEEKING system is located.

the position of Honorary Lecturer in Neurosurgery at the Royal London Hospital School of Medicine. I was delighted to follow in its great neurological tradition: in the mid-nineteenth century John Hughlings Jackson, the founding father of British neurology and neuropsychology, was a physician there. The Royal London was located in Whitechapel, an area that has been a magnet for immigrants for centuries and which therefore has always served vulnerable communities. It reminded me of Baragwanath Hospital in Soweto. It felt like a home away from home.

In the early 1990s a neurosurgical colleague in South Africa referred to me Mr S, a patient on whom, ten months previously, he had performed an operation to remove a tumour that was growing under the frontal lobes of his brain and displacing his optic nerves. Mr S had suffered a small haemorrhage during the operation, which had interrupted the blood supply to the basal forebrain (see Figure 1). Basal forebrain nuclei transmit acetylcholine to various cortical and subcortical structures involved in retrieving long-term memories. These cholinergic pathways are thought to interact with dopaminergic pathways (see Figure 2), with the latter being the so-called reward system that activates 'search' behaviours, not only in relation to physical actions in the external world but also in relation to the inner world of representations, the imaginary actions that arise in thinking and in dreams.[26] As a result of his haemorrhage, Mr S awoke from surgery with a profound amnesic syndrome, known as Korsakoff psychosis, the central feature of which is a dream-like state called confabulation. His memory for recent events was profoundly disordered in such a way that he constantly retrieved false recollections. This *search* deficit is disabling enough, but in confabulatory amnesia it is compounded by the fact that patients do not adequately *monitor* the reliability of the memories they wrongly retrieve, and therefore treat them as if they are true when they patently are not.

For example, Mr S thought he was in Johannesburg (his home town) but he had in fact just travelled to London to consult me. He had no memory of the journey. When I corrected him on that score, he insisted that he could not possibly be in London. I therefore asked him to look out of the window, since it was snowing, which never happens in Johannesburg. Initially he looked shocked, but then he composed himself and retorted: 'No, I *know* I'm in Jo'burg; just because you're eating pizza, it doesn't mean you're in Italy.'

Mr S was a fifty-six-year-old electrical engineer. I saw him in my daily out-patient clinic, six times per week, in an attempt to orientate him and help him gain some insight into the ways in which his memory was failing him. Although I saw him at the same time and place every day, he never recognised me as his therapist from one session to the next. He apparently knew my face, but routinely mistook me for someone else he knew in a different context – usually an engineering colleague who was working with him on some electronic problem, or a client seeking his professional assistance. In other words, Mr S treated me as if I were in need of his help, rather than the other way round. Another frequent misconception of his was that we were both university students, having a drink together after some sporting activity (either a rowing contest or a rugby match). I was young enough at the time for this to be plausible, but Mr S had not been a student for more than thirty years.

After each clinical session, I had a consultation with his wife in order to contextualise his misrememberings and attempt to establish their meaning. This was the main difference between the approach I was taking and the more traditional approach my colleagues took to 'cognitive rehabilitation'. Whereas neuro-psychologists conventionally concern themselves with the *degree* of memory disorder, measured from the third-person viewpoint, I was more interested in the subjective *content* of Mr S's errors, understood from the first-person perspective. I started from the assumption that the personal significance of the events that com-pulsively came to his mind, in place of the target memories that he was searching for, would cast some light on the mechanism of these confabulations – and thereby open new paths to influencing them. So, in my meetings with his wife, for example, I wanted to know whether Mr S really did belong to rowing and rugby teams when he was a student and whether he really did provide profes-sional help with electronic problems.

Two facts that I learnt in this way are relevant to under-standing his confabulations. These were, first, that he had once suffered from chronic problems with his teeth – problems which had eventually been treated (successfully) using dental implants – and, second, that he suffered from cardiac arrhythmia, which was being controlled by a pacemaker.

I have selected a short transcription from an audio recording of the first few minutes of the tenth session I had with Mr S. I have chosen this particular snippet because, when I went to collect him from the waiting room that day, he appeared briefly (for the first time) to recognise who I was and why he was consulting me. As I entered the waiting room, he touched the craniotomy scar on the top of his head and said: 'Hi, Doc.'

I was hoping to build upon this glimmer of insight, if that is what it was, as we sat down in my office.

Me: You touched your head when we met in the waiting room.

Mr S: I think the problem is that a cartridge is missing. We must … we just need the specs. What was it? A C49? Should we order it?

Me: What does a C49 cartridge do?

Mr S: Memory. It's a memory cartridge; a memory implant. But I never really understood it. In fact, I haven't used it for a good five or six months now. It seems we don't really need it. It was all chopped away by a doctor. What's his name? Dr Solms, I think. But it seems I don't really need it. The implants work fine.

Me: You are aware that something is wrong with your memory, but …

Mr S: Yes, it's not working one hundred per cent, but we don't really need it – it was just missing a few beats. The analysis showed that there was some C or Co9 missing. Denise [his first wife] brought me here to see a doctor. What's his name again? Dr Solms or something. And he did one of those heart transplant things, and now it is working fine; never misses a beat.

Me: You're aware that something is amiss. Some memories are missing, and, of course, that's worrying. You hope I can fix it, just like those other doctors fixed the problems with your teeth and your heart. But you want it so much that you are having difficulty accepting that it's not fixed already.

Mr S: Oh, I see. Yes, it's not working one hundred per cent. [He touches his head.] I got knocked on the head. Went off the field for a few minutes. But it's fine now. I suppose I shouldn't go back on. But you know me; I don't like going down. So, I asked Tim Noakes [a renowned South African sports physician] – because I've got the insurance, you know, so why not use it, why not go to the best – and he said: 'Fine, play on.'

I will break off the vignette there. It should be fairly easy to recognise the purely cognitive disorders of memory search and monitoring that I mentioned above. When Mr S saw me entering the waiting room for that tenth session, my appearance evoked a swarm of associations in him – to do with doctors, his head, missing memory, surgical procedures and the like. But in each of these instances, he did not retrieve the precise target memory he was searching for; instead, he came up with what might be characterised as near misses – memories that were in the same broad semantic categories as the targets, but which were mislocated in space and time. Thus, the idea of a 'doctor' evoked associations concerning the neurosurgeon and a famous sports physician instead of its target, me; the idea 'head' evoked a concussion incident instead of a brain tumour; 'missing memory' evoked an electronic cartridge instead of his amnesia; 'surgical procedures' evoked his earlier dental and cardiological procedures instead of the recent brain surgery, and so on. It is equally easy to see the monitoring deficit: Mr S accepted the veracity of his mistaken memories far too readily. The fact that he experienced himself as being a twenty-something student on a rugby field (despite all the evidence to the contrary) is an obvious example of this. Likewise his belief that he was still in Johannesburg.

But when Mr S's confabulations are considered from the *subjective* point of view, additional facts emerge. Imagine what it *feels* like to suddenly realise that you do not recognise the clinician who just walked into the room, although he seems to be responsible for your care; that you do not know what room (or even which city) you are in; that you have a huge scar over the top of your head, and you do not know where it comes from; that – in fact – you do not remember what happened just two minutes ago, let alone over the days and months preceding the present moment. You would probably feel something like *panic*, wondering whether this doctor might have performed an operation on your head, as a result of which you no longer remember anything from one moment to the next. This is what missing memory search and monitoring mechanisms feel like to the intentional subject of the mind – to the living I.

Now, notice what Mr S did in consequence of having these feelings (in other words, notice what causal effects they had on his cognition). Upon realising that his 'memory cartridge' was missing, he (delusionally) reassured himself that *one can simply order a new one*. Not entirely convinced by his own reassurance, he changes his mind. In fact, *one does not really need the cartridge*, one manages fine without it, and he has done so for months already. He then makes a link between the missing cartridge and the craniotomy scar: apparently something has been chopped away by a doctor. He hopes that this is not the doctor sitting before him, and moreover he hopes that the operation has not been botched. At this point, Mr S recalls that his equivalent dental and cardiological operations were successful and he (delusionally) conflates these procedures with the present one: *it was a success*, *the implants work fine* and he 'never misses a beat'. When I introduce some doubt on this score, he changes tack. He agrees that it is not working 100 per cent, but he simultaneously decides that what has happened to his head was not surgery after all, *it was merely concussion*; he is suffering the temporary effects of a minor sporting accident. Accordingly, he has been sent off the field for a few minutes. But, happily, with access to the best sports physician money can buy, he is once again reassured: *he may play on*. All will be fine.

Considering Mr S's confabulations from the first-person perspective clearly reveals something new about them: the content of his misrememberings is tendentiously *motivated*. These are far from being random search errors. They contain a clear self-serving bias; they have the aim and purpose of recasting his anxiety-ridden situation into a reassuring, safe and familiar one. So, just as Freud inferred in the case of dreams, confabulations are motivated. The mental processes in confabulatory amnesia are *wishful*. But this fact becomes apparent only when the emotional context and personal meaning (experienced by Mr S alone) of dental implants ('the implants work fine') and cardiac pacemakers ('it never misses a beat') are taken into account – as a psychoanalyst would do. This is what neuropsychologists fail to see when they aim to be entirely objective; as Sacks put it, when they exclude the psyche.

The first-person observational perspective I have just described also reveals something new about the *mechanism* of confabulation, something that is overlooked from the third-person viewpoint. It tells us that confabulation occurs not solely due to deficits in strategic search and source monitoring (i.e. missing 'memory cartridges') but also due to the release from inhibition of more *emotionally* mediated forms of recall, much as a child's memory might work. This psychodynamic mechanism has implications for the treatment of confabulation, and, of course, for the question of which brain processes are involved in it. Accurate memory search and monitoring functions turn out to depend in part upon the cholinergic basal forebrain circuits, which constrain the 'reward' mechanisms of the mesocortical-mesolimbic dopamine circuit in memory retrieval. As it happens, a similar unfettering of dopaminergic search occurs in dreams.[27] That is why I reported the case of Mr S to my colleagues under the title, 'The man who lived in a dream'.

This enabled me, as it had with dreams, to tentatively link the unconstrained dopaminergic 'reward' or 'wanting' or 'SEEKING' mechanism with Freud's notion of 'wish-fulfilment'[28] – a metapsychological concept that was closely linked with his concept of 'drive'.[29] Conversely, the functions of the cholinergic forebrain nuclei can be linked in some respects with the inhibitory influences of 'reality-testing'.[30] In this way, I began to translate Freud's inferences about the functional mechanisms of subjectivity with their physiological equivalents.

These were my first steps. Naturally, such broad generalisations cannot be based on purely clinical evidence in a single case. Having formulated my impression of Mr S, therefore, I enlisted 'blind' raters (colleagues unfamiliar with my hypothesis) to measure, on a seven-point Likert scale, the degree of pleasantness versus unpleasantness in a continuous unselected sample of 155 of his confabulations. The results were statistically (highly) significant: when compared to the target memories they replaced, Mr S's confabulations substantially improved his situation from the emotional point of view.[31] Next, my research collaborators and I demonstrated the same strong effect in studies involving numerous other patients with confabulations. In subsequent empirical studies, the mood-regulating effects of confabulation that I inferred clinically in the case of Mr S were statistically validated.[32] This programme of research opened a whole new approach to the neuropsychology of confabulation,[33] and related disorders such as anosognosia.[34] It also laid the foundations for a novel approach to common psychiatric disorders such as addiction and major depression.[35] I have spent the last three decades developing this 'neuropsychoanalytic' approach to mental illness, trying to return subjectivity to neuroscience.[36]

As I accumulated clinical experiences of the kind just described during my psychoanalytic training in London, I was invited to report my findings in a series of scientific presentations in New York. These began in 1992 with a one-day symposium at the New York Academy of Medicine, and they continued in the form of monthly seminars with my closest collaborators at the New York Psychoanalytic Society and Institute.[37] The meetings of this group of colleagues gradually spawned similar activities elsewhere in the world, resulting in the decision (in 1999) to establish a new journal to serve as a vehicle for communication between us. The journal needed a name, and so I used my invented term *Neuropsychoanalysis* for the first time.

My work in this interdisciplinary field received strong encouragement from Eric Kandel, whom I first met in 1993. Kandel is unusual among basic neuroscientists for the high regard he has for Freud. In fact, he had initially intended to train as a psychoanalyst but was dissuaded from doing so by Ernst Kris, a leading analyst of his day and the father of Kandel's then girlfriend. Kandel makes no secret of the fact that the old man thought his personality was not suited to the clinical practice of psychiatry, and I believe he is grateful that Kris guided him to brain research instead.

Five years after I first met him (and two years before he won the Nobel prize), Kandel published an article entitled 'A new intellectual framework for psychiatry' in which he argued that the psychiatry of the twenty-first century should be based upon an integration of neuroscience with psychoanalysis.[38] In a follow-up article, he said: 'Psychoanalysis still represents the most coherent and intellectually satisfying view of the mind.'[39] This opinion coincided with my own: for all its faults, psychoanalysis currently provides the best conceptual starting point for tackling subjectivity scientifically.

Not surprisingly, therefore, Kandel accepted my invitation to join the founding editorial board of *Neuropsychoanalysis*, together with a critical mass of other leading neuroscientists and psychoanalysts who agreed that this was the correct way for our disciplines to proceed.[40] One year later, as the new century dawned, we founded the International Neuropsychoanalysis Society, with Jaak Panksepp and me as its first co-chairs. This event coincided with the inaugural congress of the society, which has been held annually in different cities around the world ever since. The topic of the first congress was emotion. The meeting was held at the Royal College of Surgeons of England, with the plenary speakers being Oliver Sacks, Jaak Panksepp, Antonio Damasio and myself.

I have already mentioned my relationship with Oliver Sacks. I have also mentioned Jaak Panksepp's *Affective Neuroscience*, the title of which alluded to the fact that cognitive neuroscience paid insufficient attention to 'affect' (the technical term for feelings). Upon reading it, I established a close scientific collaboration with Panksepp, which decisively shifted the focus of my work over the next two decades away from the cortex and towards the brainstem. I am deeply indebted to him for showing the way to the insights that I will report in the pages to follow, which is why this book is dedicated to his memory.

I first became acquainted with the work of Antonio Damasio and his partner, Hanna Damasio, during my neuropsychological training. They were well-respected cognitive neuroscientists, whose textbook *Lesion Analysis in Neuropsychology* (1989) was indispensable for my dream research. What brought Damasio worldwide prominence, however, was his *Descartes' Error* (1994), an impassioned plea to cognitive neuroscience for greater recognition of affect.

Panksepp and Damasio also played a large role in the Twelfth International Neuropsychoanalysis Congress in 2011, which proved to be a turning point for the field. Held in Berlin, the topic of that congress was the embodied brain. The other plenary speakers were Bud Craig and Vittorio Gallese – two of the world's leading experts on this subject. My conventional role at these congresses is to deliver the Closing Remarks, which sum up the main themes and, most importantly, *integrate* the neuroscientific and psychoanalytic perspectives that were presented. On this occasion, my task was unusually difficult.

The first reason for the difficulty was that Damasio had clashed sharply during the congress with Craig about how a feeling 'self' is generated in the brain. Although both scientists agreed that the sense of self emerges from brain regions that monitor the state of the visceral body, Damasio – following Panksepp – took the view that the mechanisms in question were located at least partly in the brainstem. Craig, by contrast, claimed that they were located exclusively in the cortex – in the anterior insula, to be precise. This disagreement was relatively easy to resolve in my closing summary, because Damasio had provided compelling data, focusing on a patient with total obliteration of the cortical insula. I will describe this patient in the next chapter.

Far more difficult to reconcile was a major contradiction that emerged during the congress between the new views of Panksepp and Damasio on the one hand and the old ones of Freud on the other.

Damasio's patient without insular cortex 'reported feelings of hunger, thirst, and desire to void, and behaved accordingly'.[41] These feelings are examples of what Panksepp calls 'homeostatic affects' – affects that regulate the vital needs of the body. Freud called them 'drives' – the source of his 'psychical energy', the 'mainspring of the psychical mechanism'. Freud's broad term for the part of the mind that performs these vital functions was the 'id':

> The id, cut off from the external world, has a world of perception of its own. It detects with extraordinary acuteness certain changes in its interior, especially oscillations in the tension of its drive needs, and these changes become conscious as feelings in the pleasure–unpleasure series. It is hard to say, to be sure, by what means and with the help of what sensory terminal organs these perceptions come about. But it is an established fact that self-perceptions – coenaesthetic feelings and feelings of pleasure–unpleasure – govern the passage of events in the id with despotic force. The id obeys the inexorable pleasure principle.[42]

My scientific aim, you will remember, was to translate such metapsychological notions into the languages of anatomy and physiology, so that we could integrate Freud's approach into neuroscience. But here I had stumbled upon a major contradiction in Freud's classical conception: he had come to the conclusion that the 'id' was unconscious. This was one of his most fundamental conceptions about how the mind works. It was clear to me that the part of the brain that measures the 'demand upon the mind for work in consequence of its connection with the body' – the part that generates what Freud called 'drives', which are synonymous with Panksepp's 'homeostatic affects' (which trigger his wishful SEEKING mechanism) – was located in the brainstem and hypothalamus (see Figure 1). This is the part of the brain that obeys the 'pleasure principle'. But how can feelings of pleasure be unconscious? As we saw with Damasio's patient, drives such as hunger and thirst and the desire to void are *felt*. Of course they are. Yet Freud said the id – the seat of the drives – was unconscious. He had imbibed the same classical doctrine as Craig (as had I, at least initially) and had therefore located consciousness in the cerebral cortex. Thus he wrote the following in the 1920 essay that I cited previously, where he hoped that the deficiencies in his theories would vanish once we were in a position to replace psychological terms by physiological and chemical ones:

> What consciousness yields consists essentially of perceptions of excitations coming from the external world and of feelings of pleasure and unpleasure which can only arise from within the mental apparatus; it is therefore possible to assign to the system *Pcpt.-Cs.* [perceptual consciousness] a position in space. It must lie on the borderline between inside and outside; it must be turned towards the external world and must envelop the other psychical systems. It will be seen that there is nothing daringly new in these assumptions; *we have merely adopted the views on localisation held by*

cerebral anatomy, which locates the 'seat' of consciousness in the cerebral cortex – the outermost, enveloping layer of the central organ. Cerebral anatomy has no need to consider why, speaking anatomically, consciousness should be lodged on the surface of the brain instead of being safely housed somewhere in its inmost interior.[43]

In case there is any doubt about Freud's view that *all* consciousness (including feelings of pleasure and unpleasure) is located in the cortex, I will provide one more quotation from him:

The process of something becoming conscious is above all linked with the perceptions which our sense organs receive from the external world. From the topographical point of view, therefore, it is *a phenomenon which takes place in the outermost cortex of the ego.* It is true that we also receive information from the inside of the body – the feelings, which actually exercise a more peremptory influence on our mental life than external perceptions; moreover, in certain circumstances the sense organs themselves transmit feelings, sensations of pain, in addition to the perceptions specific to them. Since, however, these sensations (as we call them in contrast to conscious perceptions) also emanate from the terminal organs and since *we regard all these as prolongations or offshoots of the cortical layer*, we are still able to maintain the assertion made above. The only distinction would be that, as regards the terminal organs of sensation and feeling, the body itself would take the place of the external world.[44]

For Freud, clearly, conscious feelings, no less than perceptions, are generated in the 'ego' (the part of the mind that he identified with the cortex),[45] not in the unconscious 'id' – which I was now obliged to locate in the brainstem and hypothalamus. In short, it seemed that Freud got the functional relationship between the 'id' (brainstem) and the 'ego' (cortex) the wrong way round, at least insofar as feelings are concerned. He thought the perceiving ego was conscious and the feeling id was unconscious. Could he have got his model of the mind upside down?[46]

3

The Cortical Fallacy

In late 2004 the neuroscientist Bjorn Merker joined five families with neurologically impaired children on a week-long trip to Disney World. The children ranged in age from ten months to five years. They went on rides: a favourite was 'It's a Small World After All'. They had their pictures taken with Mickey Mouse. They ate popcorn, corn dogs and ice cream – perhaps more than was wise – and they drank sodas galore. At times they seemed to become overwhelmed; there were tears on more than a few occasions. But despite these stressful moments, what struck Dr Merker was how much the children seemed to be enjoying themselves – how they loved being there. After all, going by one of the most basic premises of neurology, these children should have been in a 'vegetative state'. They should have been one step away from coma: capable of displaying only autonomic functions like regulation of heartbeat, respiration and gastrointestinal activity, their motor responses limited to simple reflexes like blinking and swallowing.

Merker had befriended these families the year before, when he joined a self-help group for the caregivers of children suffering from the rare brain condition known as hydranencephaly. From early 2003 onwards, Merker read over 26,000 email messages passing between the group's members. On theoretical grounds, he had become concerned that the general assumption that these children will be 'vegetative' might have become a self-fulfilling prophecy, caused not by the condition itself but rather by the fact that such patients were *treated* by most paediatricians and neurologists as if they were completely insensate. Here is what he reported:

> These children are not only awake and often alert, but show responsiveness to their surroundings in the form of emotional or orienting reactions to environmental events [...] The children are, moreover, subject to the seizures of absence epilepsy. Parents recognize these lapses of accessibility in their children, commenting on them in terms such as 'she is off talking with the angels', and parents have no trouble recognizing when their child 'is back.' [...] The fact that these children exhibit such episodes would seem to be a weighty piece of evidence regarding their conscious status.[1]

Merker's most crucial observation is not that the patients lose and regain alertness but that they show 'responsiveness to their surroundings in the form of emotional or orienting reactions to environmental events'. This is the defining feature of what vegetative patients are supposed to lack: intentionality. That is why the vegetative state is also defined as '*non-responsive* wakefulness'.[2] It is generally accepted that these patients register visual, auditory and tactile stimuli *unconsciously*, but Merker saw the children expressing pleasure by smiling and laughter, and aversion by fussing, arching their backs and crying, 'their faces being animated by these emotional states'. He observed familiar adults enlisting their responsiveness to build up play sequences, predictably progressing from smiling, through giggling, to laughter and great excitement on the part of the children. They responded most vigorously to the voices and actions of their parents and other people they were familiar with, and they showed preferences for certain situations over others. For example, they appeared to enjoy specific toys, tunes or videos, and they even came to expect the regular presence of such things in the course of daily routines.

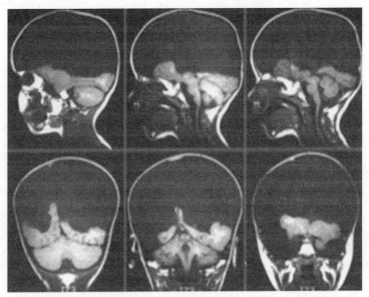

Figure 4 Brain scan of a three-year-old-girl born without a cerebral cortex. The large dark region inside the skull indicates missing tissue.

Though behaviour varied from child to child, some of them clearly showed initiative (within the limitations of their motor disabilities), for example by kicking noise-making trinkets hanging in a special frame constructed for the purpose, or by activating favourite toys using switches. Such behaviours were accompanied by situationally appropriate signs of pleasure or excitement on the part of the child.

These children clearly cannot be described as vegetative. The thing that makes all of this so astonishing is that hydranencephalic children are *born without a cortex*. This is usually due to a massive stroke in utero, which results in reabsorption of the forebrain, so that the baby's cranium is filled with cerebrospinal fluid instead of brain tissue. Hence the term 'hydranencephaly' – which means 'water instead of encephalon'. An illustration of this

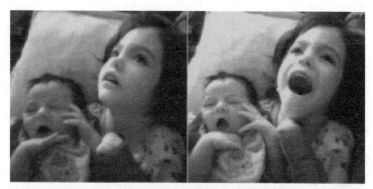

Figure 5 Reaction of a hydranencephalic girl to
her baby brother being placed in her lap.

condition is provided in Figure 4: an MRI scan of the brain of a
three-year-old girl born without a cortex. Photographs of such a
child in Figure 5 demonstrate her emotional response when her
baby brother is placed on her lap.

Where in the brain is consciousness generated? For the past 150
years, the almost universal answer to this question has been 'in
the cortex'. This was the only point on which Freud and the mainstream tradition in twentieth-century mental science could agree.
And yet, if that's right, when the cortex is absent, consciousness
ought to disappear. In the case of hydranencephalic children this
appears not to happen. All the behavioural evidence suggests that
they are, in fact, conscious. They are not comatose and nor do
they exist in a vegetative state.

Should the cortical theory of consciousness be rejected? Let's
not be too hasty. One possible objection is that the cortex was not
literally *removed*, surgically, in these children. That procedure is
called 'decortication'.

What happens when this is done? Obviously, such experimental procedures cannot be performed in human children, but
they have often been performed in other newborn mammals, such

as dogs, cats and rats. The outcome is always the same: by the objective behavioural criteria that we normally use to measure it, consciousness is *preserved*.* The post-operative behaviour of these animals cannot by any stretch of the definitions be described as 'comatose' or 'vegetative'. Merker writes that they show 'no gross abnormalities in behaviour that would allow a casual observer to identify them as impaired'. Antonio Damasio concurs: 'Decorticated mammals exhibit a remarkable persistence of coherent, goal-oriented behaviour that is consistent with feelings and consciousness.'[3] Neonatally decorticate rats, for example, stand, rear, climb, hang from bars and sleep with normal postures. They groom, play, swim, eat and defend themselves. Either sex is capable of mating successfully when paired with normal cage mates. When they grow up, the females show the essentials of maternal behaviour, which, though deficient in some respects, allow them to raise pups to maturity.[4]

The situation is even stranger than it first appears. In many respects, decorticate mammals are in fact *more* active, emotional and responsive than normal ones. Panksepp used to ask his graduate students to choose between two groups of rats, to determine from their behaviour which group had been operated upon. The students typically selected the normal ones, reasoning that the other group (really decorticate) was 'more lively'.[5]

If we are going to uphold the hypothesis that the cortex is the seat of consciousness, then these lively animals – and the expressive, emotionally responsive children that Merker observed at DisneyWorld – must in some sense be unconscious. How could this be possible?

There is a conventional answer to that question. The story goes like this. There are, in a manner of speaking, two brains, which double up on certain functions, and which talk to one

---

* Ironically, it was in part the outcome of experiments like these that eventually changed our attitudes to the ethics of such research.

another at certain points, but which are otherwise not at all alike in status. One of them (the cortex) is psychological and conscious. The other (the brainstem) is neither of those things. Information from the sense organs is fed not only to the cortex but also to the superior colliculi of the brainstem, via a set of subcortical connections (see Figure 6). These connections process sensory information, but they do it *unconsciously*. Take the famous phenomenon of 'blindsight', which occurs when the visual cortex is destroyed.[6] These patients are able to respond to visual stimuli, and yet when asked to describe what their 'seeing' is like, they report that they do not experience any visual images at all; instead, if asked, they use gut feelings to *guess* where the visual stimuli are located, which they do with remarkable accuracy. Consider the case of the patient known as TN as reported by the neuroscientist Lawrence Weiskrantz: although totally blind – in other words, entirely devoid of conscious visual experience – TN deftly manoeuvred around obstacles placed in his way along a corridor. Questioned afterwards, he reported having no idea that he was avoiding anything.[7] This surprising capacity is possible because the pathway from the optic nerve to the superior colliculi in the brainstem remained intact, despite the absence of the occipital cortex.[8]

The existence of blindsight has been taken to imply that consciousness of visual perception must happen in the cortex rather than the brainstem. It is assumed that there is 'nobody at home' in the brainstem: it is an autonomic machine that processes visual information in the same non-conscious way that a camera does. This principle applies also to the other senses, each of which involves its own specific zone of consciousness in the cortex, but each of which (apart from the sense of smell) also transmits information to the unconscious superior colliculi of the brainstem.

The little girl pictured in Figure 5 possesses no functional cortex at all. If the argument above is right, it must be that she *unconsciously* senses her baby brother being placed on her lap,

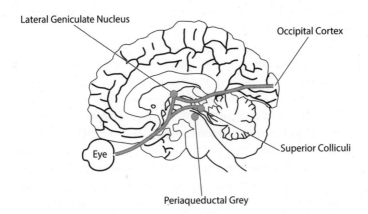

Figure 6 This figure shows the location of the superior colliculi and occipital cortex, and their connections with the eye. Similar connections exist for the other sensory modalities. Also indicated is the periaqueductal grey (PAG), which I have not yet discussed, but its relevance will become apparent shortly.

without generating any conscious awareness of the situation. In fact, she must be incapable of any sort of conscious experience whatsoever. She is something like what the philosopher David Chalmers calls a 'zombie'. Though in certain respects she acts normally, all is dark within.*

In an extensive review of the varieties of what we call 'consciousness', neurologist Adam Zeman distinguished two principal meanings of the term: 'consciousness as the waking state' and 'consciousness as experience'.[9] Anton Coenen later elaborated: 'Consciousness in the first meaning (consciousness as the waking state) is in this view a *necessary condition* for consciousness in

---

* The philosophical notion of a zombie is different from the Hollywood one: it is of an imaginary humanoid creature that acts in all respects *as if* it is conscious, and is therefore externally indistinguishable from a normal person, but it actually lacks the inner dimension of subjective experience.

the second sense (consciousness as experience or phenomenal consciousness).'[10]

The two meanings coincide with the conventional distinction in neurology between the quantitative 'level' and qualitative 'contents' of consciousness. It is therefore possible that, although decorticate animals and hydranencephalic children are awake, their experience is contentless. We have seen above that they are responsive and show behavioural initiative. Nevertheless, we can retain the hypothesis that the cortex is the seat of 'consciousness as experience' by positing that *being* conscious in the behavioural sense of being awake and responsive is significantly different from *having* consciousness in the phenomenological sense – that is, being a subject of experience.

At this stage, if you are anything like me, you might be feeling rather uneasy. The girl in Figure 5 is conscious in the sense that she is awake and responsive and initiates goal-directed movements of her own. At the same time, if 'consciousness' means having phenomenal experiences, then she supposedly doesn't have anything of the sort. To use another expression popular with philosophers, *there is not 'something it is like' to be her.*

Let me say what I find so unsettling about this line of reasoning, even if it sounds a little naïve. Judging by my own case, being awake and responsive and having conscious experience are more or less the same thing. As far as I know, I am never awake and responsive but phenomenally unconscious. The two things go together. As soon as I wake up, I become aware of things. In fact, from where I am sitting, my inner consciousness feels like it *causes* my outer responsiveness, at least to some degree. It is typically when I notice things – that is, when I become conscious of them – that I respond to them intentionally. Presumably you are the same.

That is why we gauge consciousness in neurological patients on the basis of their responsiveness. What else can we do? In clinical practice, we make distinctions between conditions like coma, the vegetative state and fully responsive wakefulness by means of the fifteen-point Glasgow Coma Scale. This scale is composed of tests of the patient's eye-opening responses, their verbal responses to questions, and their motor responses to instructions and (if necessary) to pain.[11] If the patient is fully responsive, we consider them to be conscious and we treat them accordingly. We do not worry that they might respond *as if* they were conscious while actually they are zombies. The worry goes in the other direction: neurologists must guard against the possibility that patients may be outwardly non-responsive but inwardly conscious – as occurs, for example, in cases of complete motor disability such as 'locked-in syndrome'.

Where does this leave us? It is undeniable that the philosophical problem of other minds raises doubts about what we can know about the subjective states of animals and other people from their behaviour, in much the same way as lying or play-acting can mislead us about a person's inner states. What the problem of other minds *doesn't* do is establish that the inner experience and the outer response are in fact independent from one another. Since we feel the presence of a strong link in our own individual cases, the burden of proof surely falls on those who want to claim that emotionally responsive behaviour doesn't imply phenomenal experience. This is what the rules of science normally require. This is especially true when we are dealing with human beings who, whatever other neurological disorders they suffer from, may be supposed to generate their wakeful behaviour in something like the usual way – by which I mean, through the normal functioning of the undamaged portions of their brains, rather than by lying or play-acting, or the sort of cunning contrivance one might expect to find in imaginary philosophical robots.

The implications for medical ethics are considerable. A psychiatrist colleague whose child was hydranencephalic recently shared with me a terrible dilemma she faced when she was a young mother. The neurosurgeon treating her baby suggested that an operation to close the fontanelle of the skull could be performed without an anaesthetic, since, lacking cortex, the infant was of course incapable of feeling pain. My colleague did not tell me what decision was made, and the question was too upsetting to ask. Still, you can see from cases like this one how seemingly abstract theoretical considerations might lead in short order to appalling medical errors. Let me therefore put this point forcefully: if we are to accept that someone who *seems* to be conscious actually *isn't*, we should require an extremely convincing argument. Merely raising philosophical doubt isn't enough. We need very good grounds to think that the two sorts of consciousness have come apart in such people, as they seemingly never do in us.

Currently, these grounds are supplied by the cortical theory of consciousness: the theory that consciousness arises exclusively in the cortex. This idea is accepted so routinely within behavioural neuroscience that it never occurred to me to question it, until, on the basis of my own dream research, Allen Braun directed my attention to the mysterious role played by the brainstem. On the basis of the cortical theory, doctors decide whether or not to treat living patients as sentient beings. For example, some caregivers raise hydranencephalic children in conditions of severe emotional neglect on the assumption that they are 'vegetative'. If the theory is correct, this should be perfectly fine: it isn't really neglect because the children are something like philosophical zombies. They appear to have feelings, but in reality they don't. The appearance that they do is an illusion produced by the problem of other minds. Their outer behaviour tricks us into imagining that they have any inner nature at all.[12]

What, then, is the cortical theory of consciousness, and how persuasive is it?

The first thing to notice is that it began to develop very early. The everyday observation that our consciousness consists mainly of perceptual images of events going on around us suggests that consciousness flows in through the senses. This common-sense view has no doubt been with us since we first began to think about such matters. In the seventeenth and eighteenth centuries, it gave rise to the 'empiricist' philosophies of John Locke and David Hume. They theorised that the mind – which begins as a blank slate – acquires all its specific characteristics from impressions left by sensory vibrations. The impressions were supposed to become associated with each other through regular conjunctions of various kinds[13] to produce our *memory images* of objects, which in turn became the basic building blocks of more abstract ideas. Subsequent sensory vibrations stimulate these assembled images into the forefront of consciousness, so that what we experience are not raw sensations but rather *what we have learnt* about the world.

The way in which ideas become conscious in response to an external stimulus was called *apperception* (which roughly means perceiving the present through the lens of past experience).[14] Cognitive processes like the mental imagery used in thinking were said to involve roughly the same process in reverse: internally generated activations of the memory images, suitably rearranged (and fainter than the externally generated ones).

Speculative though it was, this philosophical idea about the mind became the map that the early neurologists followed. When the nineteenth-century pioneers of modern neuropsychology sought to establish the neural correlates of these processes, they observed that the sensory organs were connected to the cortex and surmised that the sensory 'vibrations' took place in these connecting nerves. They did not, as it happens, overlook the fact that the sensory organs were connected also with subcortical nuclei. However, they assumed that the vast store of memory images which constitute our knowledge of the world must be located in the cortex, because it contains innumerably more neurons. Thus 'apperception' and the associated 'ideas' generating mental activity proper were presumed to be cortical phenomena. Theodor Meynert, the great neuroanatomical authority of the day, put it like this:

The main function of the central organ is to transmit the fact of existence to an ego gradually shaping itself in the stream of the brain [...] If we look upon the cortex as an organ functioning as a whole then the fact that it subserves the processes of the mind is all that can be said [...] To think further about the cortex is impossible and unnecessary.[15]

The cortex is, of course, impressively large in humans.[16] The neuroanatomist and neurologist Alfred Walter Campbell summarised the mainstream view of this as follows at the annual general meeting of the Medico-Psychological Association in London in 1904:

Viewed collectively, the human brain harbours two varieties of centres, controlling what we may call 'primary' and 'higher evolutionary' functions respectively; the former are those common to all animals and essential to survival, viz. centres for movement and common and special sensation; the latter are those complex psychic functions in the possession of which man rises superior to all other beings.[17]

It is important to recognise that the mind, on this view, consists entirely of memory images that reflect past experiences of the outside world. Incoming sensations merely stimulate these images and their associations into consciousness. The vibrations flowing in from the sense organs are therefore pre-mental events – *triggers* for mental activity – they are not mental events themselves. The same applies to the function of the outgoing nerves connecting the cortex with the rest of the body: there is nothing 'mental' about these pathways; they merely discharge the *outputs* of mental activity. Mental activity proper can only happen in the cortex, where the memory images reside.

Meynert described the relationship between the cortex and the outside world as follows:

The motor effects of our consciousness reacting upon the outer world are not the result of forces innate in the brain. The brain, like a fixed star, does not radiate its own heat: it obtains the energy underlying all cerebral phenomena from the world beyond it.[18]

It probably also bears pointing out that, because Freud hadn't come along yet, all mental activity was assumed to be conscious. The words 'mental' and 'conscious' were taken to mean the same thing.

The experimental work of the nineteenth century that led to our modern conception of consciousness was conducted entirely within this philosophical framework. In the late 1800s, the physiologist Hermann Munk identified the occipital cortex as the locus of the mental aspect of vision (see Figure 6). He became interested in the behaviour of dogs with experimental lesions to the occipital cortex. These unfortunate animals could see, but apparently lacked the normal 'understanding' of what they saw: they could no longer visually recognise their masters, for example, or identify their own feeding bowls, although they looked at them, and circumnavigated them, and were able to recognise them through the other senses.[19] Munk called this condition 'mind-blindness', to distinguish it from the common form of blindness that is caused by lesions of the (subcortical) sensory pathways leading from the eyes to the cortex.[20] Following empiricist philosophy, he equated what he called 'mental' vision with the capacity to activate visual *memory images* through associations, as opposed to the mechanical business of receiving raw visual sensations and firing motor reflexes. Accordingly, the condition Munk described is now called visual 'agnosia' – that is, lack of visual *knowledge*.

Clinical phenomena of exactly the same kind were soon reported in human cases – for example by the ophthalmologist Hermann Wilbrand in 1887, who described the case of Fräulein

G, a sixty-three-year-old woman who suffered a bilateral occipital stroke:[21]

> She was regarded as blind by all those around her. She was, however, quite aware that she was not completely blind, 'for when people stood at my bedside and spoke with pity of my blindness, I thought to myself: you can't really be blind because you are able to see the table-cloth over there, with the blue border, spread out on the table in the sick-room' [...] When this – in other respects highly intelligent – lady got up [after initial loss of consciousness following the stroke], she found herself in a curious state of not seeing and yet being able to see [...] Still today she is moved when she recalls her first excursion after the stroke; how absolutely different and completely strange the city appeared to be, and how extremely distressed and shaken she felt when she was led by her attendant for the first time over the Jungfernstieg and the Neuer Wall to the Stadthaus, and how the attendant indicated afresh to her the buildings and streets that were usually so familiar. She reports her reaction to the woman who was her escort at the time as being: 'If you say that that is the Jungfernstieg, and that the Neuer Wall, and this the Stadthaus, then I suppose it must indeed be so, but I do not recognise them' [...] I said to my physician: 'The conclusion may be drawn from my condition that one sees more with the brain than with the eye; the eye is but the vehicle for sight, because I see anything absolutely clearly and lucidly, but I do not recognise it, and frequently I do not know what the thing I am seeing could be.'

Wilbrand concluded that this patient suffered a loss of visual memory (i.e. a disorder of visual recognition or understanding or knowledge) rather than simple blindness. And he observed that, although blind patients can still generate visual imagery in their dreams, since their mental imagery is intact, *mind*-blind patients like Fräulein G cannot; they lose their ability to dream. How can you generate visual hallucinations without visual memory images, he asked.[22]

Such observations regarding vision were then generalised to the other modalities of perception. Thus, ablation of the auditory cortex (again in dogs) produced 'mind-deafness', now called auditory agnosia, in which the afflicted animals could no longer respond to sounds with acquired meaning, even though they clearly were not deaf: they responded to raw noise but no longer recognised their own names when called. In 1874 the neurologist Carl Wernicke observed something similar in human patients, leading to his conception of the acquired language disorders called 'aphasia'.[23]

In the motor modality, cortical lesions were said by Hugo Liepmann to likewise produce deficits of the mental aspect of movement, known as 'psychical paralysis' (or 'apraxia'). A lesion of the outgoing motor pathways caused physical paralysis, but a lesion in the cortical centre for motor memory images caused the forgetting of acquired movement skills (called 'limb-kinetic' apraxia), and a lesion of the transcortical pathways for motor associations caused a disconnection between skilled movement and abstract ideas, such as the symbolic meaning of hand gestures ('ideomotor' apraxia).[24]

Subcortical blindness, deafness and paralysis were thereby conceptualised as *physical* disorders, whereas the visual, auditory and motor effects of cortical lesions were *mental* ones: the apperceptive and associative forms of agnosia, aphasia and apraxia. The latter disorders are nowadays called 'neurocognitive' disorders. The distinction between these two classes of disorder (subcortical versus cortical) thus coincided with the disciplinary boundary between neurology and neuropsychology, as it still does.

When Munk's contemporaries ablated the entire cortex in dogs – as opposed to the specialised parts responsible for individual sensory-motor modalities – the animals did not fall into coma or a vegetative state; instead, they behaved as if they were *mindless* (as the empiricists understood the word). They became amnestic; that is, they lost all their memory images and therefore their 'understanding'. Friedrich Goltz accordingly described them as 'idiotic'. The fact that they did not fall into coma did not surprise the early investigators because, given the theoretical assumptions of their times, they expected only that decorticate dogs should lose all acquired *knowledge*. This loss need not have any effect upon the bodily sensations and reflexes, which were assumed to be subcortical. The idea that mental life consisted only in memory images was not a controversial one. The controversies of that time revolved around other things, such as how narrowly each mental function could be localised in specific regions of the cortex. Nobody doubted that all such functions were cortical.

So it was that, when Meynert summarised the whole emerging picture in his famous 1884 book *Psychiatry*,[25] he identified the mind with the *totality of memory images of objects* produced by projection of the sensory-motor periphery onto the cortex plus transcortical associations between them and those memory images that constituted abstract ideas. Naturally, he located all these images, associations and ideas in the cortex – per the quotation above. He called this the 'voluntary' portion of the brain, and he claimed it had direct connections with the body periphery via the sensory and motor nerves. That is to say, he claimed that the cortex was connected to the body *independently* of the subcortical grey matter. He acknowledged that the subcortical portion of the brain was connected with the cortex and body periphery, too, via its own separate pathways; but he described these pathways as 'reflex'. Here we have the 'two brains' I mentioned before, to explain the behaviour of hydranencephalic children.

That is how the cortex became the organ of the mind – the mind construed as *consciousness of memory images* – and how the subcortical brain became mindless. It all boiled down to the notion that your 'mind' is entirely constituted by past experience, which leaves traces that are associated with one another to form concrete images and abstract ideas. The rest of you, the peripheral sensory-motor parts and the innate and subcortical parts, including the parts that transmit impressions from the *interior* of your body, were considered to be purely reflex. And so, odd as it seems, the philosophical distinction between your mind and your body came to coincide with the anatomical distinction between the cortex and the subcortex.

In the 1960s Norman Geschwind, the great pioneer of behavioural neurology in America, enthusiastically revived these classical ideas.[26] The renaissance of empiricist associationism in neurology coincided with the 'cognitive revolution' in psychology. The information flow charts of the cognitive scientists bore a strong resemblance to the diagrams that the classical German neurologists had used to illustrate the functional relations between cortical centres containing memory images of various kinds. But – as we shall see – modern cognitivism also led to the realisation, which became firmly established by the 1990s, that many mental functions (including perception and memory) *were not conscious after all*.

Around the same time that Geschwind recommitted neuropsychology to the classical cortical theory, pioneers of what would later become affective neuroscience (people like Paul Maclean and James Olds) were accumulating observations which revealed that many subcortical nuclei connecting the brain with the interior of the body performed mental functions too. These were functions like motivational *drive* and emotional *feeling*.

Despite these parallel developments in cognitive and affective neuroscience, which I will elaborate shortly, it is ultimately on the basis of this intellectual heirloom, the attempt by nineteenth-century neurologists to confirm the theories of eighteenth-century philosophers, that today's neurologists assume hydranencephalic children to be mindless. They assume it even as they acknowledge that the old equations of 'mind' with consciousness and with cortex are no longer valid: that not all mental activity is conscious and not all consciousness is cortical. In other words, the insistence that the original equation still holds, that 'consciousness as experience' is necessarily cortical, is made on the basis of theoretical inertia rather than scientific evidence.

As you might have gathered, I don't believe that the cortical theory of consciousness is valid. In fact, I would go so far as to say that animals and human beings can be conscious even if they lack a cortex entirely. I think there *is* 'something it is like' to be a hydranencephalic child.

Of course, one of the reasons why it is so difficult to know what hydranencephalic patients experience – indeed, whether they have inner experience at all – is that they cannot speak. Language really is a cortical function, so we can't expect people who lack it to give us introspective verbal reports. They cannot provide us with the sort of subjective evidence that ordinarily makes us believe that other people are conscious, notwithstanding the problem of other minds.[27] The same applies to animals. Yet there are human beings who have lost substantial amounts of cortex who nevertheless retain the capacity to speak. In these cases, we can simply *ask* them what it's like.

The neurosurgeons Wilder Penfield and Herbert Jasper cut out large expanses of cortex (even entire hemispheres) in 750 human patients under local anaesthetic, mainly for the treatment of epilepsy. They observed that such operations have limited effects on self-reported consciousness, even in the very moment the cortex is being removed. (Brain surgery is often conducted under local anaesthetic so that the patient can report on the effects of what the surgeon is doing.) Penfield and Jasper concluded that cortical resections do not interrupt sentient being; they merely deprive patients of 'certain forms of information'.[28] I have attended many such operations myself, where my usual role is to assess the effects of electrically stimulating the memory and language parts of the cortex before the surgeon cuts there. I witnessed the very same thing that Penfield and Jasper reported.

But not all cortex is equal. It is the regions of the cortex that receive inputs from each of our special senses that generate the phenomenal 'qualia' associated with those senses (colour for vision, tone for hearing, and so on). However, according to the standard view, the raw sensations can be declared only after they have been accessed by an overarching type of consciousness that constitutes the sentient 'I'.[29]

Various terminologies are used to describe the relationship between 'phenomenal' consciousness (e.g. simple seeing and hearing) and 'reflective' consciousness (*knowing* that you are seeing and hearing). But they all convey the same basic idea: you are more than the various forms of sensory information that you process. This is why patients with damage to their primary visual cortex may be blind and those with a damaged auditory cortex may be deaf, but their sense of *self* is still there. It is this sentient self that guesses what one is seeing when one is devoid of visual qualia in blindsight, for example. These patients lose only 'certain forms of information'. So, the burning question is, what do we learn from patients who have lost those portions of the cortex that are said to be responsible for selfhood?

There are three candidate locations for this function. The first is the *insular cortex*, which is specialised for interoceptive aware-ness and is widely claimed to generate the feelings that constitute a sentient 'self'.[30] The second is the *dorsolateral prefrontal cortex*, which forms a superstructure over all other parts of the brain, and is generally believed to enable 'higher-order thoughts', including awareness of feelings.[31] The third is the *anterior cingulate cortex*, which lights up in just about every cognitive brain-imaging experi-ment, and is said to mediate things like 'self-related processing' and 'will'.[32]

Let's take these three regions one by one.

Here is an excerpt from an interview by Antonio Damasio with a patient whose insular cortex was totally obliterated by a viral illness called herpes simplex encephalitis:[33]

Q: Do you have a sense of self?
A: Yes, I do.
Q: What if I told you that you weren't here right now?
A: I'd say you've gone blind and deaf.
Q: Do you think that other people can control your thoughts?
A: No.
Q: And why do you think that's not possible?
A: You control your own mind, hopefully.
Q: What if I were to tell you that your mind was the mind of somebody else?
A: When was the transplant, I mean, the brain transplant?
Q: What if I were to tell you that I know you better than you know yourself?
A: I would think you're wrong.
Q: What if I were to tell you that you are aware that I'm aware?
A: I would say you're right.
Q: You are aware that I am aware?
A: I am aware that you are aware that I am aware.

This patient, known as 'B', was studied by Damasio's team over a period of twenty-seven years (he fell ill at forty-eight and died at seventy-three). In the published case study, they reported extensive neuropsychological data concerning his capacity to feel and respond emotionally. The main conclusions follow:

Patient B., whose insular cortices were entirely destroyed, experienced body feelings as well as emotional feelings. He reported feeling pain, pleasure, itch, tickle, happiness, sadness, apprehension, irritation, caring and compassion, and he behaved in ways consonant with such feelings when he reported experiencing them. He also reported feelings of hunger, thirst, and desire to void, and behaved accordingly. He yearned for play opportunities, for example, playing checkers, visiting with others, going for walks, and registered obvious pleasure when engaged in such activities as well as disappointment or even irritation when the opportunities were denied. [...] Given the impoverishment of his imagination, Patient B.'s existence was a virtually continuous 'affective' reaction to his own body states and to the modest demands posed by the world around him, undampened by high-order cognitive controls [...] According to Craig (2011), the tell-tale sign of self-awareness is the ability to recognize oneself in a mirror, an ability that in his words 'can only be provided by a functional, emotionally valid neural representation of self'. Patient B. passed this test consistently and repeatedly. In brief, these findings run counter to the proposal that human self-awareness, along with the ability to feel, would depend entirely on the insular cortices and, specifically, on its anterior third (Craig 2009, 2011). In the absence of insular cortices, we need to entertain neuroanatomical alternatives to explain the basis for B.'s feeling abilities and sentience.[34]

Bud Craig and others claim that the affective feelings which constitute the self are generated in the insular cortex. And yet Damasio found that Patient B was, like the decorticate rats, actually *more* emotional after his cortical lesion than he was prior to it. The prediction that such patients should lose their sentient 'presence' is clearly disconfirmed.

The same occurs in human patients with damage to the second region under consideration here: the *prefrontal* cortex, which forms a superstructure over all other parts of the brain. The emotionality of prefrontal patients is widely recognised; it forms a core feature of the 'frontal lobe syndrome'. This had been observed already in 1868, in the famous case of Phineas Gage, after an iron bar passed through his skull in an accident at work:

> The equilibrium or balance, so to speak, between his intellectual faculties and animal propensities, seems to have been destroyed. He is fitful, irreverent, indulging at times in the grossest profanity (which was not previously his custom), manifesting but little deference for his fellows, impatient of restraint or advice when it conflicts with his desires.[35]

Some leading neuroscientists of emotion, such as Joseph LeDoux, believe that feelings literally *come into existence* in the dorsolateral prefrontal cortex.[36] On this view, the subcortical precursors of feelings are entirely unconscious until they are 'labelled' in conscious working memory.[37] For these theorists, emotion is just another form of cognition – in fact a rather abstract, reflective form of cognition. But if they are right, then why do patients whose working memory is largely obliterated show so much emotion? Phineas Gage is only the most famous case in a substantial literature. What might reflective consciousness add to the picture, if the pre-reflective form of feeling is so vividly expressed in these patients' behaviour?

To be fair, most cortical theorists are not especially interested in feelings. They focus instead on the contribution made by the prefrontal lobes to higher-order *thought*. Hypotheses arising from these broader theories, too, are readily testable in human cases with damage to the prefrontal lobes. Complete prefrontal lesions are very rare – but they are not unheard of.

My patient 'W' was forty-eight years old when I first examined him. He scored full marks on the Glasgow Coma Scale, meaning that he was totally awake and responsive. At the age of thirteen he had suffered a burst aneurysm in his brain, which required an operation in which the frontal lobes were retracted to wrap the vulnerable blood vessel and so prevent further bleeding. That surgery unfortunately led to a chronic brain infection, requiring multiple additional operations, which ultimately resulted in the total destruction of the prefrontal lobes on both sides (see Figure 7).

Fortunately, his language cortex was spared. Here is part of our conversation:

Q: Are you consciously aware of your thoughts?
A: Yes, of course I am.
Q: In order to confirm that, I am going to ask you to solve a problem that will require you to consciously picture a situation in your mind.
A: OK.
Q: Imagine that you have two dogs and one chicken.
A: OK.
Q: Do you see them in your mind's eye?
A: Yes.
Q: Now tell me how many legs do you see, in total?
A: Eight.
Q: Eight?
A: Yes; the dogs ate the chicken.

The punchline was delivered with a mischievous grin. It might not be the best joke in the world, but it provides convincing evidence that, colloquially speaking, someone was home. Patient W said he was conscious, as did Damasio's Patient B, and in the absence of any reason more compelling than radical philosophical scepticism, I am inclined to take their words for it.

The third and last cortical region that is supposed to have a special relationship with the sentient 'I' is the *anterior cingulate gyrus*. Patients with complete bilateral lesions in this region are relatively easy to come by, partly because this part of the cortex

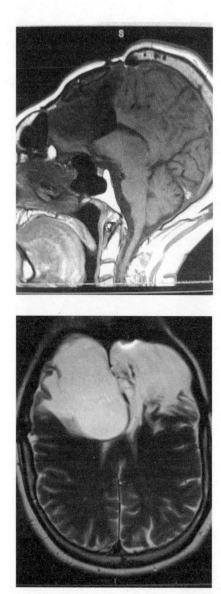

Figure 7 MRI scan of the brain of Patient W, showing complete destruction of the prefrontal lobes bilaterally.

has been a regular target of surgery for psychiatric conditions, such as obsessive-compulsive disorder. (The anterior cingulate region has brought this attention upon its head, as it were, precisely because of its association with 'self-related processing'.)

In the acute post-operative period some of these patients experience a breakdown of the distinction between fantasy and reality. Charles Whitty and Walpole Lewin described a striking example (their Case 1):

> In response to a question as to what he had been doing that day, he replied: 'I have been having tea with my wife.' Without further comment from the questioner, he continued: 'Oh, I haven't really. She's not been here today. But as soon as I close my eyes the scene occurs so vividly. I can see the cups and saucers and hear her pouring out. Then just as I lift the cup for a drink I wake up, and there's no one there.'
> Q. Do you sleep a lot, then?
> A. No, I am not asleep – it's a sort of waking dream [...] even with my eyes open sometimes [...] My thoughts seem to be out of control, they go off on their own – so vivid. I am not sure half the time if I just thought it or it really happened.[38]

Note the frequent reference to 'I'. Whitty and Lewin suspected their patient might have been suffering complex seizures, but similar experiences have been reported in many other cases with bilateral lesions of the anterior cingulate gyrus. Here, for example, is one of my own.[39] She was a forty-four-year-old woman who suffered a subarachnoid brain haemorrhage. She reported the following:

> It's as if my thinking becomes real – as I think about something, so I see it actually happening before my eyes, and then I also become very confused and I don't know sometimes what has really happened and what I am just thinking.

She gave an example:

Patient: I was lying in my bed thinking, and then it sort of just happened that my [deceased] husband was there talking to me. And then I went and bathed the children, and then, all of a sudden, I opened my eyes and 'Where am I?' – and I am alone!
Me: Had you fallen asleep?
Patient: I don't think so; it's as if my thoughts just turned into reality.

Consciousness is manifestly abnormal in these cases, but that is not the point. Nobody is disputing that the cortex is involved in conscious processing. What is unsupportable is the view that the conscious self is *generated* there.

We have seen lively, intentional, responsive and emotional behaviour from humans and animals who lack cortex altogether. Introspection suggests that consciousness is intimately involved in these behaviours in our own (presumptively normal) cases; so, the problem of other minds notwithstanding, we would need a good reason to think it isn't involved in these cases as well. But when we go in search of those good reasons, we find only the dead weight of academic history, a pre-neurological model of the mind that happened to be used as a template by the brain's early explorers. Very few scientists today would endorse Hume's version of empiricism or Meynert's *Psychiatry*. Indeed, Meynert's views are widely derided as 'brain mythology'.[40] And yet the dogma that the cortex is the organ of the mind has become the grounding assumption of an entire field of medicine.

The cortex is, of course, involved in many cognitive functions, including language. We therefore can't hope for introspective verbal reports from people in whom it is entirely absent. But we have first-hand testimony from patients whose impairments allow for self-reflective verbal declarations, even though they lack the specific parts of the cortex that are supposed to give rise to overarching self-awareness. Time after time, these patients claim to be conscious and declare their introspective 'being'. Are they lying? If so, why? And what on earth could it mean to say that a person without a self is lying about their selfhood? Cognitive neuroscience is teetering on the brink of incoherence here; a good sign that it has taken a wrong turn.

As far as the evidence goes, in my view, the cortical theory is untenable. There is no good reason to believe that the cortex gives rise to sentient existence, in the sense that you and I ordinarily experience it, and many good reasons to conclude that it does not. We will have to look for the source of our being elsewhere.

4

What is Experienced?

Do you have to be aware of what you are perceiving and learning in order to perceive and learn it? Common sense, perhaps, says yes. The empiricist philosophers said yes. But the answer is no.

The ideas of the empiricists gave rise to the classical neuroanatomical view in which perception (and the memory traces derived from it) is the basic ingredient of consciousness. But the scientific evidence showing that *we are unaware of most of what we perceive and learn* is now overwhelming. Perception and memory are not inherently conscious brain functions. In this respect, common sense was wrong. It turns out that everything your mind does (except one thing, as we shall see) can be done pretty well unconsciously.

This insight, strangely enough, is attributable to Sigmund Freud.

Despite his lifelong support for the cortical theory of consciousness, Freud was one of the first neuroscientists to question his teacher Meynert's distinction between the 'mental' and 'non-mental' parts of the brain.[1] After examining the available data, he concluded in 1891 that the views of 'Munk and other researchers who base themselves on Meynert [...] can no longer be maintained'.[2] Specifically, he found no clear anatomical distinction between Meynert's 'voluntary' and 'reflex' pathways. He also showed (by counting the number of nerve fibres involved) that cortical memory images were only generated after a series of intermediate links between them and the sensory periphery.

These links were located in the supposedly *non-mental* parts of the brain, and they reduced the amount of information being transmitted at each stage. Freud inferred that these subcortical links must be doing something to the sensory information they are processing. Note his intriguing metaphor, conceived in a pre-digital age:

> We know only that the fibres which reach the cerebral cortex after their progression through [the subcortical] grey tissues still maintain some relationship to the periphery of the body, but they can no longer deliver an image which resembles it topologically. They contain the body periphery in the same way as [...] a poem contains the alphabet, in a complete rearrangement, serving different purposes, with manifold links between the individual topological elements, whereby some of them may be rendered several times, others not at all.[3]

If the cortex is not connected directly with the body periphery, but rather via intermediate subcortical links, then the memory images deposited in the cortex cannot be literal projections of the world outside. They must be the end products of multi-stage information processing. Since this processing culminates in cortical memory images, the subcortical part of the processing must in some sense generate *preliminary versions* of the memory images. The subcortical links must therefore provide part of the mental processing that we call 'apperception'. It makes no sense, Freud argued, to draw an artificial line between the subcortical and cortical parts of the processing and claim that only the final product is 'mental':

> Is it justifiable to take a nerve fibre, which over the whole length of its course has been only a physiological structure and subject to physiological modifications, and to immerse

its end in the mind and to furnish this end with an idea or a memory image?[4]

Since Freud assumed, like everyone else did in those days, that only cortical processes are *conscious*, this led him to entertain the notion that perception and learning must include *unconscious* preliminary stages which are just as 'mental' as the conscious ones. That is, unconscious (subcortical) memory traces must be just as mental as cortical ones; they too are part of the function we call 'memory', even though they lack consciousness. Freud concluded in a letter to Wilhelm Fliess: 'What is essentially new about my theory is the thesis that memory is present not once but several times over, that it is laid down in various kinds of indications.'[5]

He identified five successive stages in the processing of information: 'perception', 'perceptual trace', 'unconscious', 'preconscious' and 'conscious' (see Figure 8). The crucial difference between the stages was not that the unconscious one was bodily and the preconscious one was mental, but rather that the preconscious kind of memory processing could be replayed in consciousness while the unconscious kind could not.[6] In other words, *only a part of the mind is conscious*.

In 1895 Freud made a similar point about the nervous pathways leading from the *interior* of the body. Here too, he argued, it makes no sense to claim that visceral bodily information becomes 'mental' only when it reaches the cortex. Since this information (which conveys feelings such as hunger and thirst) makes demands upon the mind for work in consequence of its connection with the body, we must accommodate these demands, too, within our picture of the mind. Even if you are not aware of your biological 'drives', as Freud called them, they surely form part of your mind. Recall from Chapter 2 that Freud went so far as to suggest that the demands of the body provide 'the mainspring of the psychical mechanism' and that the forebrain (the cortex of which represents the outside world) is merely a 'sympathetic ganglion'. In

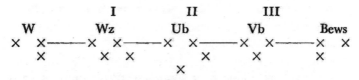

Figure 8 Freud's first diagram of the memory systems, which accompanied his letter to Wilhelm Fliess. (W = Perception, Wz = Perceptual Trace, Ub = Unconscious, Vb = Preconscious, *Bews* = Conscious.)

other words, he concluded that both conscious and unconscious memory images are formed 'in sympathy' with the demands of the body – that we only perceptually represent and learn about the outside world because we must meet our biological needs there.

To this view of the brain, modern science has added many details of which Freud had no inkling, and we have corrected him on some important points. All the same, his basic conclusion is generally accepted in neuroscience today: the brain performs a wide range of *mental* functions that do not enter consciousness. The title of a famous review of the relevant literature by the modern cognitive scientist John Kihlstrom says it all: there is indeed 'Perception without awareness of what is perceived, learning without awareness of what is learned'.[7] Another well-known review by John Bargh and Tanya Chartrand summarises our current understanding under a more poetic title, 'The unbearable automaticity of being':

Most of moment-to-moment psychological life must occur through non-conscious means if it is to occur at all [...] To consciously and wilfully regulate one's own behaviour, evaluations, decisions, and emotional states requires considerable effort and is relatively slow. Moreover, it appears to require a limited resource that is quickly used up, so conscious self-regulatory acts can only occur sparingly and for a short time. On the other hand, the non-conscious or automatic [psychological processes ...] are unintended, effortless, very fast, and many of them can operate at any given time. Most important, they are effortless, continually in gear guiding the individual safely through the day.[8]

However, our modern recognition of the ubiquity of unconscious mental functioning was not based on Freud's theories. It was arrived at entirely independently, on the basis of fresh neurological findings and experimental evidence. A turning point was the case of HM, which was published in 1957. As a result of surgical resection of his hippocampi (the part of the cortex responsible for encoding declarative memories, the part causing him to suffer epileptic seizures), he was completely unable to remember any events that occurred after the date of his operation. However, the neuropsychologists who investigated him repeatedly from then (1953) until his death in 2008 observed that his performances on psychometric tests improved dramatically over the years, showing clear practice effects. He learnt how to master the tests even though he had no conscious recollection of doing them.[9]

Similar observations were made earlier by Édouard Claparède, although he didn't recognise their significance. He reported the case of an amnesic woman who shook his hand every day as if she had never met him before (like my patient, Mr S). Claparède decided to hide a drawing pin in his palm, to see if she would remain so eager to greet him after being pricked. Following this experience, she refused to shake his hand, even though she had no conscious recollection of the painful event. When asked why she refused, she offered only evasions, such as: 'Does a lady not have the right to withhold her hand from a gentleman?'[10]

Equivalent unconscious processing was demonstrated for apperception itself. In the days before institutional review boards vetted the ethics of medical research protocols, Roger Sperry reported the case of a woman whose cortical hemispheres were surgically separated to control intractable epilepsy. When pornographic images were flashed to her isolated right hemisphere, she giggled and blushed, even though her (verbal) left hemisphere could not consciously 'declare' what the feelings were about. She could offer only tangential statements to explain her embarrassment, saying things like 'That is some machine you've got there, Dr Sperry!'[11]

What is important to notice here is that unconscious mental processing is not restricted to subcortical brain structures. In Sperry's case, for example, the patient was unable to bring to awareness the embarrassing perceptual information that her right cerebral cortex had received, processed and recognised.

The fact that the cortex can perform such functions without awareness was conclusively demonstrated by many experimental studies. In one of them, negative and positive words were flashed very briefly to research participants – so briefly that they were unaware of having seen anything at all. Their subsequent behaviour was clearly influenced by the words they claimed not to have seen: for example, after negative adjectives were flashed in association with a photograph of face A and positive ones with face B, the participants showed a preference for face B, although they did not know why they preferred it.[12] This shows that the negative and positive words must have been seen, read *and understood* unconsciously. Since reading with comprehension is an exclusively cortical function – a function of precisely the kind that the classical anatomists considered quintessentially 'mental' – we can only conclude that *cortical functions are not inherently conscious.*

This is the generally accepted view in cognitive science today: the mind construed as memory images (now called 'representations') is not intrinsically conscious. The part of the brain that generates the 'contents' of consciousness ('consciousness as experience') – the cortex – can do the very same thing in the absence of conscious experience. But if that is the case then what makes its unconscious functions *mental*? What makes the cortex – the supposed organ of the mind – different from other information-processing devices like mobile phones? Once again, we are teetering on the brink of incoherence.

This brings us back to the big question with which I concluded Chapter 3: if the cortex is not where our consciousness comes from, then where *does* it come from? And, moreover, if most 'moment-to-moment psychological life' carries on without conscious experience, then why does it *ever* involve conscious experience? Why doesn't all this information processing go on non-consciously?

The history of cognitive neuroscience is a graveyard of theories that tried to specify the function of consciousness. What they have in common is the assumption that its function is to 'bind' the multitude of information streams that are spread throughout the brain into the coherent whole that characterises our conscious experience. For example, reading and face recognition and colour perception and movement perception and object recognition and space perception and the like, all take place simultaneously in widely distributed parts of the brain. How do they come together in the unified visual images that we ordinarily perceive – with the colour and movement of a face, for example, happening in just the right locations?

Meynert claimed that this binding function was performed by the transcortical fibres, which 'associate' memory images with each other. More than a hundred years later, we have not made much progress beyond that hypothesis. James Newman and Bernard Baars proposed that a unified 'global workspace' is generated in the cortex by the thalamus, rendering the disparate bits of information globally accessible to experience. Stanislas Dehaene and Lionel Naccache added that the prefrontal and parietal association cortical areas integrate the activities of the primary sensory zones into this workspace. Gerald Edelman introduced thalamocortical 're-entrant loops' as the key function, whereby the integrated information is sent back to earlier levels of perceptual processing. Giulio Tononi emphasised the resultant 'massively integrated' information processing, claiming that it is the degree of integration between the bits that is the key component; consciousness is a function of *how much* information is integrated. Francis Crick and Christof Koch hypothesised that the synchrony of gamma oscillations in the cortex binds and stores the experiences – in other words, that the integration might happen in time rather than space. Rodolfo Llinas similarly suggested synchronisation of thalamocortical activity below 40Hz. And so on.[13]

None of these theories tells us why or how the binding of information – through association, oscillation, synchronisation, re-entry, massive integration and the like – should *necessarily* give rise to experience. Why, exactly, should the contents of a 'global workspace' for information processing be experienced consciously? Does not binding and storage and synchronisation and massive integration and so on occur also in *unconscious* processing of information? Computers generate global workspaces and massively integrate information all the time, when they are linked together by the internet. Why, then, should the internet not be conscious?

A worrying development, arising directly from these contemporary theories, is the fact that an increasing number of respectable neuroscientists (such as Koch and Tononi) now suggest that it *might* be. This is because their theories oblige them to accept this strange possibility. In doing so, they are following the 'pan-psychist' turn initiated by Thomas Nagel, according to whom *all* things might be (just a little bit) conscious.[14]

We should evaluate these theories in light of the observations I reported earlier, to the effect that consciousness persists in the absence of cortex. Newman and Baars locate their 'global workspace' in the cerebral cortex, and yet decorticate animals and children born without a cortex appear to be conscious. At the very least, the evidence that these animals and children are sentient beings is a lot more compelling than the evidence that the internet is. Moreover, speaking adults in whom the supposedly crucial substrate for massively integrated information processing has been destroyed by disease tell us that their sense of self persists without that substrate. This applied to my Patient W, whose prefrontal lobes were obliterated completely.

None of this should surprise us if we combine such clinical observations with the experimental evidence I reviewed earlier, to the effect that the cortex does most of its information processing (like reading and face recognition) *unconsciously*.

There is something remarkable buried in the evidence I have presented so far which I would now like to bring to the fore. When Claparède's patient refused to shake his hand, she did not know why. Yet she presumably *felt* some aversion to doing so. She must have had some subjective basis for refusing his hand, even though she had no access to the objective cause (the memory of the drawing pin). The same applies to Sperry's patient: she felt embarrassed but did not know why. In other words, she was unconscious of the objective cause of her feeling (the pornographic images) but not of the subjective feelings that accompanied them (her embarrassment).

The same thing probably applied in the word-flashing experiment: the participants must have felt some preference for face B, even though they lacked awareness of the reason why. This seems also to have happened with the hydranencephalic child shown in Figure 5: she displayed subjective pleasure when her baby brother was placed in her lap, even though she could not possibly have known the objective cause of her feeling, since she was entirely devoid of cortical images.

Therefore, in all these instances, when I said the patients and research participants I described were 'unconscious' of the causes of their actions, I spoke too loosely. They were unconscious of certain perceptions or memories – representations – but their *feeling* persisted. There was still 'something it is like' to be them, making their value judgements. They were conscious of their feelings; they were unconscious only of where the feelings came from.

This gives us a major clue as to what sentience fundamentally consists in. Apparently alone among mental functions, feeling is *necessarily* conscious. Who ever heard of a feeling that has no subjective quality? What would the point be of a feeling if you did not feel it? Even Freud accepted that feelings *must* be conscious:

> It is surely of the essence of an emotion that we should be aware of it, i.e., that it should become known to consciousness. Thus the possibility of the attribute of unconsciousness would be completely excluded as far as emotions, feelings and affects are concerned.[15]

I am aware that some people (including psychoanalysts) disagree with this statement and claim that unconscious emotions do exist. I will return to this matter shortly. For now, let me be absolutely clear about what I mean by the term 'feeling': I mean that aspect of an emotion (or any affect) that you *feel*. I mean the feeling itself. If you do not feel something, it is not a feeling.

A few paragraphs back I said that Claparède's patient 'presumably' felt some aversion, Sperry's patient 'must have' felt embarrassment, the word-flashing test participants 'probably' felt a preference, and so on. I did so because neuroscientists typically do not ask about subjective experiences. That is the only reason why we don't know what these research subjects felt. By contrast, I (like Freud and Sacks) take my patients' introspective reports very seriously. In this way, I hope to avoid errors such as the conflation of dreaming with REM sleep.

If a hundred out of a hundred people report a feeling of pain when you pinch their hands, it is not unreasonable to conclude that pinching people's hands causes pain, even if we must rely upon introspective reports of it in each instance. This is even more true for observations we can replicate in our own cases, where we have direct access to the phenomenon in question. If pinching my hand causes me to feel pain, then I have personal experience of what others mean when they report 'pain'.

Feelings are difficult to research because they are inherently subjective, but (*pace* the behaviourists) that doesn't give us a licence to ignore them. If we exclude feelings from our account of the brain, we will never understand how it works. The role that feelings played in Mr S's confabulations illustrates this point very well.

When we look at what turns up in our own consciousness, we notice a few general categories of content. There are, of course, 'representations' of the outside world – both perceptions and memories of it and thoughts about it. Philosophers have paid a lot of attention to these representations, and the empiricist model of the mind was designed to account for them. But they are not the only thing we find in consciousness.

There are also feelings – feelings about what is going on in the world, feelings about our thoughts about that world, feelings (above all) about ourselves, including feelings that seem to be reports on the condition of our bodies. There are free-floating feelings as well, the emotions and moods that qualify our experience of the world and shape our behaviour within it. Sometimes they register as bodily sensations; still, many moods seem attributable neither to the condition of our bodies nor to anything we can put our fingers on in the world outside. Isn't consciousness full of feelings like this? And yet, to an amazing degree, neuroscientists searching for an explanation of consciousness have ignored them.[16]

The extent to which the empiricist philosophers and their scientific heirs, the behaviourists and cognitive scientists, ignored feeling is astonishing.[17] The behaviourists asserted that all learning is governed by 'rewards' and 'punishments', and yet they never told us what those things are. They conducted rigorous experiments, which gave rise to the 'Law of Effect'. This law states that if a behaviour is consistently followed by rewards it will increase, and if it is consistently followed by punishments it will decrease. This process of learning from experience was called 'conditioning'.

Edward Thorndike, the author of the Law of Effect, wanted to prove that animals learn by trial and error, not by thinking. But Thorndike's law actually amounts to a Law of *Affect*,[18] since it implies that behaviours which make us (and other animals) feel good are the behaviours that we repeat, and those which make us feel bad are the ones we avoid. The Law of Effect is in its essence, therefore, nothing other than Freud's 'pleasure principle'. But the behaviourists couldn't accept the existence of anything as subjective as feeling. B. F. Skinner, for example, notoriously declared that: 'the "emotions" are excellent examples of the fictional causes to which we commonly attribute behaviour'.[19]

Thorndike's original wording of his law reveals the lie: 'Responses that produce a satisfying effect in a particular situation become more likely to occur again in that situation, and responses that produce a discomforting effect become less likely to occur again in that situation.'[20] The words 'satisfying' and 'discomforting' were subsequently edited out and replaced by 'reinforcing' and 'punishing'. Here's why:

> The new terms, 'reinforcing' and 'punishing' are used differently in psychology than they are colloquially. Something that reinforces a behaviour makes it more likely that that behaviour will occur again, and something that punishes a behaviour makes it less likely that that behaviour will occur again.[21]

This definition of the terms is completely empty, as it has to be, since the pivotal word is 'something'. The same applies to the behaviourist terms 'positive' and 'negative' reinforcement:

what is it that makes a reinforcement positive or negative if not a feeling?

The intended meaning of 'reward' and 'punishment' seems to be that the value inheres *in the stimulus* – rather than in the recipient of the stimulus. If a horse approaches me and I give it a sugar lump, it will (by the Law of Effect) be more likely to approach me again, whereas if I squirt a lemon in its face, it will be less likely to do so. According to Thorndike, the sugar lump and the lemon *themselves* thereby become rewarding or punishing of the horse's behaviour; there is no need to consider the feelings they evoke, if such things even exist. This is, of course, faulty reasoning which mislocates the causal force, resulting in the problem that Sacks complained of: a mind stripped of agency. It's one thing to treat the brain as a black box for methodological reasons. It's another thing entirely to attribute causal powers to things outside the box that they don't in fact possess, and then conclude that nothing is happening in the box.

The idea that feelings are 'fictional' has had many baneful effects upon science. For example, during most of the last century, when the basic physiological mechanisms of energy balance were being researched, behaviourists forbade use of the words 'hunger' and 'satiation', since such things cannot be seen or touched. Behavioural scientists would only speak of the 'incentives' and 'rewards' associated with feeding behaviour. This is not just a semantic issue. If a word like 'hunger' cannot be used, then how can we understand the role it plays in regulating eating? Might this, in turn, not retard the development of hunger-reducing treatments for obesity? In fact, feeding behaviour is regulated by two interacting brain mechanisms: a 'homeostatic' system, which regulates energy stores, and a 'hedonic' system, which mediates *appetite*.[22] And just as with bodily affects like hunger, might not the prohibition of emotional words like 'sadness' and 'fear' delay the development of antidepressant and anti-anxiety treatments?[23] If feelings really do exist, then surely they have physiological correlates which will be overlooked if we do not take them into account.

In what follows I am going to proceed differently from the behaviourists. Following Panksepp, who was the first neuroscientist with the temerity to use words like 'hunger' to explain the regulation of energy balance in both humans and other animals, I will accept that feelings *do* exist.[24] I will assume that you know what it is like to feel thirsty, or sad, or sleepy, or amused, or confident, or uncertain. This assumption is no less justified than other scientific inferences about things in nature. And it can be tested in the usual way, through falsifiable predictions.

Feelings are real, and we know about them because they permeate our consciousness. They are, in fact, for the reasons I will now explain, the wellspring of sentient being – in a sense that seems to me barely metaphorical. From their origin in some of the most ancient strata of the brain, they irrigate the dead soil of unconscious representations and bring them to mental life.

5

Feelings

To the delight of his doctors and of us all, my brother Lee recovered well from his brain injury and made an excellent adjustment. Though he could be easily led astray by people seeking to take advantage of him, he was physically large and strong and he used that to good effect. He was never quite the same, but he got on with life.

I found it harder to adjust.

When I was five years old, our father bought Lee a wristwatch, which he promised to give him when he learnt to tell the time. A brief lesson followed. Lee struggled with questions about what it meant when the short hand was at 9 and the long hand at 11. Observing from across the room, I blurted out the answers. My father told me to be quiet. That evening he gave Lee the watch, even though he hadn't quite mastered it yet. Where's my watch? I thought.

As I grew older, such childish feelings gave way to guilt. I was doing well at school. Lee was lucky to finish it. Things just seemed to come to me more easily than they did to him. I wished my family could talk about this problem, and find strategies to avoid my causing him distress. My mother, however, couldn't bear any discussion of the accident, and my father was not the sort of person with whom I could discuss *anything* heartfelt – least of all how he governed the family.

In retrospect it seems obvious, but when I made the decision to study neuropsychology, I had no inkling that it might have anything to do with Lee's accident. I *do* remember thinking as a child that the only thing worth doing in life was to find out what 'being' is. Yet it was only during my psychoanalytic training, years later, when I was already working as a neuropsychologist, that I joined the dots. Clearly, my career choice was a compromise between ambition on the one hand and guilt on the other. If I became a neuropsychologist, I could be academically successful and at the same time help people in difficulty. That presumably also explained why I was so frustrated by purely academic neuropsychology, and why I decided to become a clinician, and even an occasional neurorehabilitation therapist – professional roles that tend to be looked down upon by neuroscientists.

Let me pause the reminiscence for a moment. Do you find this a plausible account of my decision? It isn't a trick question: I think that my submerged memories of these experiences explain why I made the career choices that I did. In general, though, isn't this often how our feelings seem to motivate us, somewhere below the threshold of our awareness? It's a commonplace of psychological explanation – a piece of Freudianism so thoroughly absorbed into ordinary common sense that few would think to object to it. How odd, then, to assert that feelings are conscious by definition: that they are in some sense the essence of consciousness. How could that be?

One day in 1985, after completing my undergraduate studies, I set off to Baragwanath Hospital for my first day as an intern neuropsychologist. The psychosomatic backdrop was that I was worried: my professor – Michael Saling – had departed for sabbatical leave in Australia, and I was going to be without the usual supervision. My stress was multiplied by the fact that I got lost on the way to the hospital, in Soweto, a township that was extremely dangerous for white boys like me.

When I finally found the hospital, relieved to have made it in one piece, I rushed to Ward 7 (Neurosurgery) and asked for Mr Percy Miller, the consultant neurosurgeon to whom I had been instructed to report. The matron led me through the ward. There was a strong smell: a mixture of bodily fluids, antiseptic and something like boiled cabbage.

Miller was a birdlike man. Standing over a patient while drawing cerebrospinal fluid from a portal in his head, he engaged me intensely. 'Are you Mark Solms? Good to meet you. Mike told us you're standing in for him. Glad you're here. Just wait. I'll finish up and take you on a ward round.'

I tried to look calm and professional. Standing in for him? The neurosurgeon washed his hands and then took me to the first bedside. Before I registered what was happening, he rattled off the patient's clinical history. 'This is Mr So-and-So, forty-three years old, left temporal astrocytoma, grade three. We are operating on Wednesday. Get us some baseline measures of his language functions. Memory too, I suppose; you know better than me.'

I wondered whether to ask what a grade three astrocytoma was.

We moved to the next bed. 'This is Mr So-and-So, fifty-eight. He has a pituitary macroadenoma. We're doing him tomorrow; so please assess him first. We'll take a transsphenoidal approach. Mike finds the cognitive outcomes better – which makes sense.'

So it went, from bed to bed. By the time we got to the fifth or sixth one, it was too late to confess that I didn't understand most of what was being said. Also, I couldn't remember the important points about the first few cases. I could hardly remember anything. I told myself I would go back and look at the files later.

We pushed through a pair of swing doors into the female ward. The surgeon kept talking: 'This is Ms So-and-So, thirty-six years; cysticercosis.' I didn't know what that was either. The next patient was stark naked. Some of the men had been naked, too, but this was different. I shouldn't be seeing this, I thought. Do they know I'm not a doctor? As we proceeded, I noticed the patients' facial expressions. One looked at me plaintively, another blankly, some anxiously, many not at all.

I was completely out of my depth. These people assumed I knew what I was doing. And because I was going along with the assumption, it was quite clear that I was going to be responsible for someone's death.

The paediatric ward was worse still. The first child was a case of rampant hydrocephalus: his head was like ET's, twice the normal size. I felt sick and hot and was perspiring profusely. The next child had a soft balloon of flesh protruding from the back of her neck. She was staring at the wall. Percy Miller said something about a 'myelomeningocele'. His voice receded and was replaced by the slow thumping of my heart. My vision tunnelled, and, as his talking mouth faded from view, I had one last thought: I am going to hit my head on the floor, crack my skull open, and end up in one of these beds. Feeling oddly relieved, I blacked out.

This example of my first day at work illustrates many of the essential features of 'affect', including the fact that, although we don't always know why we feel things, we certainly know *what* we feel. In this instance, I felt impending doom, which I linked intellectually with my professional incompetence, behind which lay all sorts of complicated feelings about my brother, including both my need to help him and a deep identification with him. But the immediate cause of my fainting was a primitive *bodily* reflex, triggered directly by my emotions.

Vasovagal syncope causes you to faint because your brain reacts to something alarming, usually the sight of blood or some other perceived risk of physical injury. This trigger (registered by the amygdala) activates the solitary nucleus in your brainstem, which causes your heart rate and blood pressure to drop suddenly. That in turn leads to reduced perfusion of your brain; and you lose consciousness.

Why do we have this innate reflex? The answer is: it reduces blood flow and thereby staunches haemorrhaging, in anticipation of injury.[1] It is only in us humans that the reflex causes fainting, due to our upright posture and large brains, which requires more cardiac effort. (The theory that fainting is a form of 'playing dead' is therefore unlikely.) On balance, the vasovagal reflex enhances rather than reduces your chances of surviving bodily injury; and, anyway, cerebral blood flow is restored as soon as you fall to the ground.

Let's take another example.[2] Respiratory control is normally automatic: so long as the levels of oxygen and carbon dioxide in your blood stay within viable bounds, you don't have to be aware of your breathing in order to breathe. When blood gases exceed these normal limits, however, respiratory control intrudes upon consciousness in the form of an acute feeling called 'air hunger'. Unexpected blood gas values are an indication that *action is required*. It is urgently necessary to remove an airway obstruction or to get out of a carbon-dioxide-filled room. At this point, respiratory control enters your consciousness, via an inner warning system that we experience as *alarm* – specifically, in this case, suffocation alarm.

The simplest forms of feeling – hunger, thirst, sleepiness, muscle fatigue, nausea, coldness, urinary urgency, the need to defecate and the like – might not seem like affects, but that is what they are. What distinguishes affective states from other mental states is that they are hedonically *valenced*: they feel 'good' or 'bad'. This is how affective sensations such as hunger and thirst differ from sensory ones like vision and hearing. Sight and sound do not possess intrinsic value – but feeling does.[3]

The goodness or badness of a feeling tells you something about the state of the biological need that lies behind it. Thus, thirst feels bad and quenching it feels good, because it is necessary to maintain your hydration within the bounds that are viable for your survival. The same applies to the unpleasant feeling of hunger in relation to the pleasurable relief that is brought about by eating. In short, pleasure and unpleasure tell you *how you are doing* in relation to your biological needs. Valence reflects the value system underwriting all biological life, namely that it is 'good' to survive and to reproduce and 'bad' not to do so.

What motivates each individual is not these biological values directly, of course, but rather the subjective feelings they give rise to – even if we have no inkling of what the underlying biological values are, and even if we do not intellectually endorse them. For example, we eat sweet things because they taste good, not because they tend to have high energy content, which is the biological reason why they taste good. Affects tell long evolutionary stories of which we are completely unaware. As in the case of the vasovagal reflex, we are conscious only of the feelings.

The reason I use the word 'unpleasure' rather than 'pain' is that there are many different kinds of pleasure and unpleasure in the brain. Hunger feels bad, and it feels good to relieve it by eating; a distended bowel feels bad, and it feels good to relieve it by defecating; pain feels bad, and it feels good to withdraw from the source of it. These are bodily affects but the same applies to emotional ones. Separation distress feels bad and we respond to it by seeking reunion. Fear feels bad and we escape it by fleeing the danger (and sometimes by fainting). Suffocation alarm and hunger and sleepiness and fear all feel bad, but they feel bad in different ways. Getting rid of them, by contrast, feels good, also in different ways.

The different feelings signal different situations of biological significance, and each one compels us to *do* something different: urinating cannot satisfy hunger and eating cannot relieve a full bladder. Recall what Damasio said of Patient B: 'He reported feelings of hunger, thirst, and desire to void, *and behaved accordingly*'. To do otherwise would be fatal.

Feelings make creatures like us do something *necessary*. In that sense, they are measures of demands for work. With air hunger, the required work must rebalance your blood gases. With hypothermia, the work must return you to a viable temperature range. With separation distress, it must reunite you with a caregiver. And so on. In the jargon of control theory, blood gas imbalances, temperature undershoots, missing caregivers and approaching predators are 'error signals', and the actions they give rise to are meant to *correct* the errors. The resolution of affect through something like satiation means that an error has been successfully corrected, whereafter it disappears from the radar of consciousness.

We are, once again, in territory that will seem oddly familiar to psychoanalysts. Freud, you will recall, defined 'drive' as 'a measure of the demand made upon the mind for work in consequence of its connection with the body'. Now you see how closely affects are tied to drives; they are their subjective manifestation. Affects are how we become aware of our drives; they tell us how well or badly things are going in relation to the specific needs they measure.

This is what affects are for: they convey *which* biological things are going well or badly for us, and they arouse us to do something about them. In this respect, affective sensations are different from perceptual ones. Philosophers are often exercised by the possibility of what they call 'qualia inversion'. How do I know that the red I see is the same as the red you see? Mightn't my red look blue to you? The problem of other minds suggests that we can never know, because you and I would both point to the same objects in the world and call them 'red'. But what is true of visual perception is not true of affective experience. Redness does not *cause* anything different than blueness, so you can arbitrarily swap them around without any physical consequences.

The same does not apply to feelings like hunger in relation to urinary urgency, or fear in relation to separation distress. One feeling (fear) impels you to escape something; another (separation distress) impels you to search for someone. The feeling is inextricable from the bodily state it entrains.[4] If you swapped them around, you would feel an irresistible urge to escape from a missing caregiver and you would tearfully search for a stalking predator. If you swapped subjective redness with blueness there would be no consequences, but if you swapped the feeling of fear with separation distress (or hunger with urinary urgency), it would kill you.[5]

The second thing to notice about feelings is that *they are always conscious*. A feeling of which you are unconscious is not a feeling. This is true by definition, as I said earlier, but it is also true because of a specific feature of the brain's physiology. We will find out why that is so in the next chapter, when we discuss the brain mechanisms of 'arousal'. For now, I want to convince you that feelings are always conscious, without exception. That is not to say that all the need-regulating mechanisms in the brain are conscious, but that is my point: *it makes a difference whether a need is felt or not*. Your water-to-salt ratio may be sliding all the time, in the background, but when you feel it, you want to drink. You might objectively be in danger without noticing it, but when you feel it you look for ways to escape.

Different things call for different names, and the difference between felt and unfelt needs makes it necessary to introduce a terminological distinction. 'Needs' are different from 'affects'. Bodily needs can be registered and regulated *autonomically*, as in the examples of cardiovascular and respiratory control, thermoregulation and glucose metabolism. These are called 'vegetative' functions, and with a good reason: there is nothing conscious about them. Hence the term autonomic 'reflex'. Consciousness enters the equation only when needs are felt. This is when they make demands on *you* for work. (Please note: drives measure demands made upon *the mind* for work; some needs never arouse voluntary action and therefore never become conscious. Blood pressure is a clinically notorious example; you know nothing about the under and overshoots until it is too late.)

Emotional needs, too, can be managed automatically, by means of behavioural stereotypes such as 'instincts' (inborn survival and reproductive strategies, which Freud placed at the centre of his conception of the unconscious mind). But emotional needs are usually more difficult to satisfy than bodily ones, for reasons I will explain later in this chapter. That is why the feelings they evoke are typically more sustained. A feeling disappears from consciousness when the need it announces has been met.

The third thing to notice about feelings is that felt needs are *prioritised* over unfelt ones. We are constantly beset by multiple needs. Vegetative functions like energy balance, respiratory control, digestion, thermoregulation and the like are going on constantly; and so are stereotyped behaviours of various kinds. You could not possibly feel all these things simultaneously, not least because you can only *do* one thing (or very few things) at a time. A selection must be made. This is done on a *contextual* basis. Priorities are determined by the *relative* strengths of your needs (the size of the error signals) *in relation to* the range of opportunities afforded by your current circumstances. Here is a simple example. While giving a lecture, my bladder gradually distends, but I do not feel the increasing pressure until the lecture is over, at which point I suddenly need to pee. I *become aware* of the error signal because of the contextual change. At that point, sensing the opportunity, my bodily need becomes conscious as a feeling.[6]

Prioritising needs in this way has major consequences. The most important one is that *when you become aware of a need, when it is felt, it governs your voluntary behaviour.*

What does 'voluntary' mean? It means the opposite of 'automatic'. It means subject to here-and-now *choices*. Choices can be made only if they are grounded in a value system – the thing that determines 'goodness' versus 'badness'. Otherwise, your responses to unfamiliar events would be random. This brings us full circle, back to the most fundamental feature of affect: its valence. You decide what to do and what not to do on the basis of *the felt consequences of your actions*. This is the Law of Affect. Voluntary behaviour, guided by affect, thereby bestows an enormous adaptive advantage over involuntary behaviour: it liberates us from the shackles of automaticity and enables us to survive in unpredicted situations.

The fact that voluntary behaviour must be conscious reveals the deepest biological function of feeling: it guides our behaviour in conditions of *uncertainty*. It enables us to determine in the heat of the moment whether one course of action is better or worse than another. In the example of air hunger, the regulation of your blood gases becomes conscious when you don't have a ready-made solution to maintain your physiologically viable bounds. In your rush to escape from the carbon-dioxide-filled room, for instance, how do you know where to turn? You have never been in this situation before (in any burning building, let alone this specific one) so you cannot possibly predict what to do. Now you must decide whether to go this way or that, up or down, etc. You make such decisions by *feeling your way through the problem*: the feeling of suffocation waxes or wanes, depending upon whether you are going the right way or not – that is, depending upon whether the availability of oxygen increases or decreases.

The conscious feeling of suffocation alarm involves different neural structures and chemicals from those responsible for unconscious respiratory control, just as the feeling of hunger recruits different brain systems from those responsible for autonomic regulation of energy balance. Science would never have discovered these things if it had continued to ignore feelings.

There is a lot more to be said about feelings, such as how they enable *learning from experience* (through the Law of Affect) and how they relate to thinking, but the points I have just made, in a nutshell, explain *why* we feel. I have described what feeling adds to the repertoire of mechanisms we living creatures use to stay alive and reproduce. This is what psychology contributes to biology. Natural selection determined these survival mechanisms, but once feelings evolved – that is, the unique ability we have as complex organisms to *register our own states* – something utterly new appeared in the universe: subjective being.

It's hard to imagine how one could be in physical pain without feeling it: to be in pain just is to feel it. But what about the subtler affective states we call emotions and moods? Does one have to feel that one is happy to be happy? Isn't it quite common to belatedly realise that you are in the grip of a bad mood – to come to the conclusion that you must be depressed, and so on?

We have started exploring affect via its bodily forms. This is because they provide the simplest examples, and no doubt they were also the first to appear in evolution. I think the 'dawn of consciousness' involved nothing more elaborate than valenced somatic sensations. What I want to show you now is that human emotions are complex versions of the same type of thing. They, too, are ultimately 'error' signals which register deviations from your biologically preferred states, which tell you whether the steps you are taking are making things better or worse for you.

There is unfortunately no generally agreed upon classification of affects in current neuropsychology. I have drawn a distinction between bodily and emotional affects, but such sharp demarcations do not exist in nature. In drawing this line, I am following Jaak Panksepp's taxonomy, which is widely – but not universally – accepted. He further divided bodily affects, of which there are a great variety, into interoceptive ('homeostatic') and exteroceptive ('sensory') subtypes. Hunger and thirst, for example, are homeostatic affects, whereas pain, surprise and disgust are sensory ones.* So, to be clear, according to Panksepp there are three types of affect: homeostatic and sensory ones (both of which are bodily) and emotional ones (which involve the body but cannot be described as 'bodily' in any simple sense). Think, for example, of missing your brother, which is an emotional state; it is not bodily in the same way that hunger and pain are.

---

* Panksepp (1998). I do not use Panksepp's term 'homeostatic' for the interoceptive subtype of bodily affect because, as we shall soon see, all affects are homeostatic. His narrower use of the term can therefore be confusing.

Panksepp based his taxonomy upon *deep brain stimulation* studies which he and his students, and many others before them, performed on thousands of animals. I visited his laboratory on many occasions and it was a veritable menagerie, at different times filled with pigeons, chickens, beagles, guinea pigs, rats and prairie voles. (The illness that finally killed him might well have been caused by excessive exposure to some of these animals, especially the birds.)

There can be no doubt that Jaak felt compassion, even love, for these animals. But one has also to acknowledge that they were sacrificed to science, in numbers, without themselves choosing that fate. It is a sad irony that we owe to Panksepp's research the almost certain knowledge that the animals listed above are sentient creatures, subject to intense emotions, which are in their essence no different from those that you and I feel. As a result of these findings, and the increasing concern about the ethical issues that flow from them, Panksepp spent the last decades of his life studying only positive emotions.

In what follows, when I switch back and forth from observations about animals to observations about humans, I am doing so deliberately. As Panksepp said when he was accused by colleagues of anthropomorphism towards animals: he would rather plead guilty to zoomorphism towards humans. The purpose of his experiments was to determine which brain structures and circuits reliably arouse the same affective responses, not only across individuals but also across species. When it comes to emotional affects, it turned out that seven of them can be reliably reproduced not only in all primates but also *in all mammals*, by stimulation of exactly the same brain structures and chemicals. (Many of them can be evoked in birds, too, and some in all vertebrates.) Mammals separated from birds about 200 million years ago; that is how old these emotions are. Still, since humans are mammals, in what follows I am going to focus on these seven types. As far as we know, these are the *basic* ingredients of the entire human emotional repertoire. All our myriad joys and sorrows appear to be the outputs of these seven systems, blending with each other and with higher cognitive processes.

Perhaps the best-known alternative taxonomy of the 'basic emotions' (as they are called) is that of Paul Ekman.[7] The disparities arise primarily from the fact that Ekman used a different method from Panksepp, namely the study of facial expressions and related behaviours. As Charles Darwin observed long ago in *The Expression of Emotions in Man and Animals* (1872), these show remarkable commonalities across mammal species. However, although Panksepp and Ekman classify the affects differently (for example, Ekman considers disgust to be an 'emotional' affect whereas Panksepp calls it a 'sensory' one), there is wide agreement about the affects themselves. Nobody seriously doubts that disgust exists, so in a sense it doesn't matter all that much how we classify it.

The major dissenting voice is that of Lisa Feldman Barrett. Again, the disagreement is attributable mainly to methodological differences.[8] She focuses on self-reported emotions in humans and, not surprisingly, finds that there is enormous variability in how different people (and cultures) characterise and parse feelings. This does not disprove the fact that basic natural kinds lurk beneath the socially constructed surface. I will soon illustrate the mechanisms whereby such variability arises, but the short explanation is this: our reflexes and instincts provide rough-and-ready tools for survival and reproductive success, but they cannot possibly equip us adequately for the multiplicity of unpredicted situations and environments that we find ourselves in. We therefore need to adaptively supplement the innate responses through learning from experience. The fact that human beings do so with such ease is the major reason why, for better or worse, we came to dominate the world to the degree that we now do.

The instinctual programmes that undergird actions in humans are typically so conditioned through learning that they are no longer recognisable as 'instinctual'. Yet instincts and reflexes are always there in the background. The whole of psychoanalytic theory rests upon this insight: if you take the trouble to find them, implicit instinctual tendencies can always be discerned behind explicit intentions.

Here comes a quick introduction to the instinctual emotions in Panksepp's taxonomy.[9] Just as different scientists classify affects differently, so they use different words for them. Panksepp capitalised his terms for the basic emotions to distinguish them from colloquial usage – that is, to indicate that he was talking about whole brain functions, not only the feelings.

(1) LUST. We are not constantly sexually aroused. Erotic feeling enters consciousness only when sex is prioritised over other motives, which happens in the context of fluctuating needs and opportunities. When sexual desire is aroused, you feel it, then erotic feelings guide your actions. You pay attention to different details when you are sexually aroused compared to when you are fearful, for example, and you behave differently too. In this way, your exteroceptive consciousness and voluntary behaviour is determined by your inner state; you experience the world differently – literally bringing different experiences upon your head – depending on what you are feeling. That is also why you cannot easily become simultaneously sexually attracted and fearfully repulsed; you cannot prioritise them both. When the need for safety is prioritised, sexual motives are driven from your mind.

It is uncertain whether LUST should be classified as a 'bodily' or an 'emotional' affect. Some people even doubt that sexuality is a need. This is an excellent example of the difference between (unconscious) needs and the (conscious) affects they give rise to. When we engage in sexual acts, we are not usually trying to perform our biological duty. In fact, very frequently, we are hoping *not* to reproduce. As with sweet tastes versus energy supplies, what motivates us subjective beings is the pursuit of erotic pleasure, not reproductive success. That is, we are driven by *feelings*. But living organisms *need* to reproduce, at least on average. That is why sex became subjectively pleasurable in the first place, through natural selection.

I say 'on average' because not all sexual activity results in reproduction, only enough of it to keep the species going. This exemplifies another central principle: the limited utility of inborn behaviours to meet our emotional needs. In sex, the inborn aspects boil down to little more than genital engorgement and lubrication, lordosis (arching the back, which makes the vagina available for penetration), mounting, intromission, thrusting and ejaculation. Together with these reflexes, stroking the clitoris or penis (which are equivalent organs) at a certain rhythm produces pleasurable sensations which predict the release of sexual tension, ultimately through orgasm to satiation. These involuntary contrivances do not equip us for the difficult task of persuading other people – especially the particular ones we are attracted to – to comply with our desire to have sex with them. The main reason why 'emotional' needs are more difficult to meet than 'bodily' ones is that they typically involve other sentient agents, who have needs of their own; they are not mere substances like food and water. To satisfy sexual needs, therefore, we must supplement our innate knowledge with other skills, acquired through *learning*. This fact alone explains the wide variety of sexual activities that we indulge in, alongside the 'average' form that was bequeathed by natural selection.

Notice that learning does not *erase* reflexes and instincts; it elaborates, supplements and overrules them, but they are still there. Street lamps illuminate pathways by night, but they cannot get rid of the darkness altogether. The usual mechanism for updating long-term memories, 'reconsolidation' (which I'll describe in Chapter 10), doesn't apply to reflexes and instincts. That is because reflexes and instincts are not *memories*; they are fundamental dispositions that are 'hard-wired' into each species through natural selection.

Our range of sexual behaviours is increased further by the fact that the brain circuitry for both female-typical and male-typical LUST exists in every mammal. The tendency that comes to dominate is determined by various factors, including genetic and intrauterine events.[10] I will not go into the anatomical and chemical details here, except to point out that both circuits arise in the hypothalamus and terminate in the periaqueductal grey, abbreviated PAG. (The location of the PAG was shown in Figure 6. You will learn shortly why it is so important.) In other words, they are entirely subcortical.[11]

(2) SEEKING. All the bodily needs (and sexual ones) – which are registered by 'need detectors' located mainly in the medial hypothalamus – activate this second emotional drive, which was introduced in Chapter 1. It is almost synonymous with Freud's 'libido' concept, but Freud did not know that LUST merely *activates* this system; LUST and SEEKING are not the same thing. SEEKING generates exploratory 'foraging' behaviour, accompanied by a conscious feeling state that may be characterised as expectancy, interest, curiosity, enthusiasm or optimism. Think of a dog in an open field: no matter what its current bodily needs are, foraging propels it to engage positively with the environment, so that it might satisfy them there. Almost everything that we living creatures need is 'out there'; through foraging we learn, almost accidentally, what things in the world satisfy each of our needs. In this way, we encode their cause-and-effect relations. This illustrates again how stereotyped instincts lead to individualised learning.

SEEKING is unusual among the basic emotions in that it *proactively* engages with uncertainty. This is the origin of novelty-seeking and even risk-taking behaviours. Foraging makes us explore interesting things, so that we will know what to expect when we encounter them in future. Once a dog has explored a hedge, for example, and familiarised itself with its contents, it will be less interested in it the next time round. Accordingly, SEEKING is our 'default' emotion. When we are not in the grip of one of the other ('task-related') affects, our consciousness tends towards this generalised sense of interest in the world.

Anatomically, the neurons of the SEEKING circuit arise from the ventral tegmental area of the brainstem, from where they course upwards, via the lateral hypothalamus to the nucleus accumbens, amygdala and frontal cortex (see Figure 2). Chemically, its command neuromodulator is dopamine, the 'stuff of dreams' (see Chapter 1).[12] This reveals an interesting fact about SEEKING, namely that it can be aroused even during sleep by demands made upon the mind for work, leading to problem-solving activities which must be guided by conscious feelings. Hence we dream.

(3) RAGE. While we engage positively with the world through SEEKING, in the optimistic belief that our needs will be met there, things do not always go well for us. Just as evolutionary prehistory equipped us with reflexes and instincts that reliably predict ways to meet our bodily needs, so too we are born with emotional tendencies that predict ways to get us out of trouble. In challenging situations of universal significance, appropriate feelings are prioritised to govern behaviour. We are thereby spared the biological costs of having to reinvent the wheels that enabled our ancestors to survive and reproduce. Emotions are a precious inheritance. They transmit innate survival skills – implicit, unconscious knowledge – in the conscious form of feelings that can explicitly guide our actions.

When the RAGE system is triggered – as it is by anything that gets between us and whatever could otherwise meet our current needs – our consciousness is qualified by feelings ranging from irritated frustration to blind fury. The reflexes and instincts then released include piloerection (hair standing on end), protrusion of nails, hissing, growling and baring of teeth, followed by 'affective attack': lunging at the target of your wrath and biting, kicking or hitting it, until it relents.

Why do you *feel* the affects that accompany such behaviour? The answer is as before: the feelings tell you how you are doing, whether things are going well or badly, as you try to rid yourself of an obstacle – one that is often simultaneously trying to get rid of you. You sense the sweet taste of victory or the bitterness of defeat. This guides what you do next, including the possibility that pain (a sensory affect which is suppressed during affective attack) might become prioritised, thereby replacing RAGE and putting an end to the fight – and perhaps leading to flight.

How could this all go on automatically, without ongoing conscious evaluation? This question applies all the more to the role of affect in *thinking* – a topic that we might fruitfully introduce now. Thinking is 'virtual' action; the capacity to try things out in imagination; a capacity which, for obvious biological reasons, saves lives. This capacity is not unique to humans, but it is particularly highly developed in us. So, let's turn to a human story. Picture this situation, derived from my own experience. My headmaster is berating me in his office. I feel increasingly angry. The instinctual response is affective attack. Now I *think* about the potential consequences. Instead of lunging at him, therefore, I inhibit the instinctual action tendency and *imagine* my range of alternatives; I feel my way through them. Eventually, I settle upon a satisfactory solution: after leaving his office, when nobody is looking, I deflate his car's tyres. In this way, I reduce my RAGE without suffering dire consequences. This illustrates, once again, why innate behavioural stereotypes must be supplemented by learning from experience, including the imaginary form of experience called thinking. When faced with real-life frustrations, which frequently include *conflicting* needs (in this case, RAGE versus FEAR), instinctual solutions are not enough. But again, please notice: supplementing instinctual responses through learning does not erase them. I decided not to attack my headmaster, but the inclination to do so remained, and would arise again in similar situations in future. (This is not a uniquely human story. A dog might not have come up with a solution like this, but primates pull all sorts of cunning tricks.)[13]

Emotions like RAGE are not 'mere' feelings. Emotions play a fundamental role in survival. Imagine the consequences if we didn't stake claims on the available resources and prevent others from taking our share. If we couldn't become frustrated, irritated or angry, we wouldn't be inclined to fight for what we need; in which case, sooner or later, we'd be dead. It is easy to overlook the biological function of emotion in the civilised conditions under which we live today. But we have only been living like this (in permanent settlements with artificial laws regulating social behaviour) for about 12,000 years. Civilisation is a very recent feature of mammalian existence; it played no part in the design of our brains.

Conscious thinking requires cortex. But the feelings that guide it do not. The circuit mediating RAGE is almost entirely subcortical, and, like all the other affective circuits, its final destination is the brainstem PAG.[14]

(4) FEAR. The fight/flight dichotomy shows that affective attack is not always the best way to deal with an adversary. The contextual factors separating fight from flight are encoded in the amygdala, which mediates both RAGE and FEAR.[15]

Most mammals 'know' from day one that some things are inherently frightening. Newborn rodents, for example, freeze when exposed to a single cat hair – although they have never experienced cats and know nothing of their attitude to mice. It is easy to see why this is so; if each mouse had to learn from experience how to respond to cats, that would be the end of mice. Again, we see the enormous biological value of emotions.

We humans fear dangers like heights, dark places and creatures that slither and crawl towards us, and we avoid them by the same instincts and reflexes as other mammals: freezing and fleeing behaviours. Unlike the vasovagal shut-down reflex discussed previously, these 'escape' behaviours are facilitated by rapid breathing, increased heart rate and redirection of blood from the gut to the skeletal musculature. (Hence the loss of bowel control associated with extreme fear.) As with other emotions, the conscious feeling of fear tells you whether you are heading towards or away from safety.

An interesting example is provided by Patient SM, who was in her late twenties when her case was first published. She suffered from Urbach-Wiethe disease, a rare genetic condition that results in bilateral calcification of the amygdala. She felt no fear. In consequence:

> [She] has been the victim of numerous acts of crime and traumatic and life-threatening encounters. She has been held up at both knifepoint and gunpoint, was almost killed in a domestic violence incident, and has received explicit death threats on multiple occasions. Despite the life-threatening nature of many of these situations, S.M. did not exhibit any signs of desperation, urgency, or other behavioural responses that would normally be associated with such incidents. The disproportionate number of traumatic events in S.M.'s life has been attributed to [...] a marked impairment on her part of detecting looming threats in her environment and learning to steer clear of potentially dangerous situations.[16]

I have studied a large number of patients like this, as there is an unusually high incidence of Urbach-Wiethe disease in a remote corner of South Africa called Namaqualand, which happens to be near my birthplace. (The faulty gene was carried there by a German colonist and then concentrated in an isolated community.) I found their dreams particularly interesting; they are short, simple and manifestly wishful. One of the patients I studied, whose husband was unemployed, dreamt, 'My husband found a job; I was very happy.' Another, who was the mother of a disabled child, dreamt, 'My daughter could walk; I was very happy.' Yet another, whose father had died, dreamt, 'I saw my father again; I was so happy.' These dreams are typical of Urbach-Wiethe sufferers, whose fearless imaginations expect no dangers in meeting their desires.[17]

Most people seem to be born with some specific FEAR triggers. Can you imagine what would happen if each of us had to learn from experience what happens when you jump off a cliff or pick up a viper? That is why we are descended from ancestors who felt disinclined to try. Those who did try are not our ancestors because they left no offspring. We have every reason to be grateful for this inheritance.

But then we must learn *what else* to fear. We learn from experience – including thinking – that things other than falling from dizzy heights and snake bites can cause similar harm. Electrical sockets and the shocks they produce, for example, could not have been predicted by evolution, but they are just as dangerous as vipers. Also, we must learn what else *to do* when we are fearful, to supplement our instinctual responses. It is not adaptive to freeze before or flee from everything that scares you, just as it is not adaptive to attack everyone who frustrates you. It should be clear by now what role conscious feelings play in this learning process; they tell you what works and what doesn't, before it's too late, and thereby help you stay alive.

Fear conditioning reveals important additional facts about what is conscious and what is not. One of its special features is 'single-exposure learning'; you need only stick your finger into an electric socket *once* to prevent you from ever doing it again. It is easy to see why; you were lucky to survive the first time, so why repeat the experience? However, as with all the other biological mechanisms that underwrite emotions, you don't have to know this in any 'declarative' sense; conditioning just happens automatically. This is because FEAR conditioning does not require the involvement of the cortex. It can occur in early childhood, even before the hippocampus (the cortical structure responsible for laying down declarative long-term memories) has matured. For this reason, just like Claparède's case, many neurologically healthy people fear things without knowing why.

Cognitive scientists attribute this to 'unconscious' learning, but as already noted, that is only because they neglect affect. It is true that many people are unconscious of the reasons they fear things, but they are only too conscious of the associated feelings. Feeling is all that is required to guide voluntary behaviour. Take the word-flashing experiment I described earlier: if words like 'murderer' and 'rapist' are subliminally associated with face A, and 'caring' and 'generous' with face B, research subjects will feel a preference when they are subsequently required to choose between them, even though they cannot say why. 'Gut feeling' is what guides this choice, but feelings easily go unrecognised; so they are described cognitively by words like 'guessing'.[18]

This explains much of the perplexity that surrounds 'unconscious emotions' in cognitive science. It is not the *emotions* that are unconscious so much as the *cognitive* things they are about. As we saw above in relation to thinking, it can certainly help to know what your feelings are about, but that insight is not essential. In fact, sometimes it is better *not* to think before you act, not least because thinking takes time.

The same applies to fear-conditioning. Once you have learnt to fear something – especially if you do not consciously know why – the association is well-nigh irreversible. As Joseph LeDoux memorably put it, fear memories are 'indelible'.[19] This reveals important facts about unconscious memory in general, which I shall outline later. For now, I will mention only that 'non-declarative' memories (like emotional and procedural ones) are hard to forget, for the same reason that they're unconscious: they entail less uncertainty (i.e. they are more generalisable) and are therefore less subject to contextual revision. This is how acquired behaviours become stereotyped and automatised. Insofar as the purpose of cognition is to learn how to meet your needs in the world, automatisation is the ideal of learning.

(5) PANIC/GRIEF. Separation anxiety is different from FEAR. It emerges initially in infancy, when you become instinctually attached to your mother (or main caregiver). Unlike fear conditioning, but for equally good biological reasons, this takes about six months: one instance of nurturant care is not enough to show that someone can be relied upon forever.

When mammals become separated from their attachment figures, a stereotyped sequence unfolds, starting with 'protest' behaviour and followed by 'despair'. The protest phase is characterised by feelings of panic, together with distress vocalisations and searching behaviour. The panic is frequently combined with anger – 'where *is* she?' – which evokes another conflict, this time between PANIC/GRIEF and RAGE. The one emotion makes you want to keep your caregiver close to you, always and forever, whereas the other simultaneously makes you want to destroy her. Guilt, a *secondary* emotion which inhibits RAGE, is the typical learnt outcome. This is a good example of how secondary emotions (like guilt, shame, envy and jealousy) arise from conflictual situations. Unlike the basic emotions, these are learnt constructs – hybrids of emotion and cognition (as Barrett's research shows).

The despair phase is characterised by feelings of hopelessness – literally 'giving up'. The standard explanation is: if the separated pup's crying and searching do not lead quickly to reunion, then the potential costs of alerting predators to its vulnerable state begin to outweigh the benefits. Also, if the pup wanders too far from home base, its chances of being found when its mother returns are reduced. Thus, on statistical balance, as occurred with vasovagal shutdown, giving up (despite how painful it is) becomes the inherited survival strategy.

Here is a classical description of the separation cascade in human children, by the psychoanalyst John Bowlby:

> [Protest …] may begin immediately or may be delayed; it lasts from a few hours to a week or more. During it the young child appears acutely distressed at having lost his mother and seeks to recapture her by the full exercise of his limited resources. He will often cry loudly, shake his cot, throw himself about, and look eagerly towards any sight or sound which might prove to be his missing mother. All his behaviour suggests strong expectation that she will return. Meanwhile he is apt to reject all alternative figures who offer to do things for him, though some children will cling desperately to a nurse.
>
> [Despair …] succeeds protest, the child's preoccupation with his missing mother is still evident, though his behaviour suggests increasing hopelessness. The active physical movements diminish or come to an end, and he may cry monotonously or intermittently. He is withdrawn and inactive, makes no demands on people in the environment, and appears to be in a state of deep mourning.[20]

The latter state is, of course, akin to depression, which is often accompanied by guilt. Accordingly, Panksepp and others (including me) applied his elucidation of the brain mechanisms of PANIC/GRIEF to developing new treatments for mood disorders.[21] Chemically, the transition from 'protest' to 'despair' is mediated by peptides called opioids, which shut down dopamine (for the effects of which, see the case described below, on p. 123). That is why depression is characterised by the mirror opposites of the feelings that characterise SEEKING.[22] The anatomical trajectory of the PANIC component of this system descends from the anterior cingulate gyrus to the PAG, which is where all the emotion circuits terminate.[23] (Later I will explain how it happens that all affective cycles both end *and* begin in the brainstem PAG.)

It is interesting that this opioid-mediated circuit evolved from the brain's older analgesic system; the mental anguish of loss is an elaboration of the bodily mechanisms for sensory pain.[24] This is a good example of the seamless transition that exists in nature between life-saving sensory affects and emotional feelings. There is nothing 'fictitious' about emotions. The painful feelings associated with separation and loss – coupled with learning from experience – play a causal role in ensuring the survival of mammals and birds, which *need* caregivers. This applies beyond childhood, too: the brain circuits just described mediate attachment bonds throughout life, as they do, sadly, mediate many other forms of addiction.

(6) CARE is the other side of attachment; we not only need loving care ourselves, we also need to look after little ones, especially our own offspring. The so-called maternal instinct exists in all of us, but not to the same degree, because it is mediated by chemicals found at higher levels (on the average) in females: oestrogen, prolactin, progesterone and oxytocin – all of which rise dramatically during pregnancy and childbirth.[25] Also noteworthy is the overlap between the brain chemistry and circuitry for CARE, PANIC/GRIEF and female-typical LUST.[26] These facts alone could explain why depression is so much more common (almost three times) in women than men. Approximately 80 per cent of human females somehow know from childhood that it is 'good' to cradle babies to the left of the body midline, whereas males tend to discover this (instinctually) after they father children.[27] On the other hand, even completely inexperienced boys usually know what to do when a baby cries. They do not prod it with their fingers or pick it up by the foot to see if that helps; they know (they innately predict) that a 'good' thing to do is to hold it close and rock it while making soothing noises.

And yet, as every parent learns, this is not enough. Successfully raising an infant to maturity requires a lot more than instinct. Therefore, as with the other emotions, we must learn from experience what to do in the myriad unpredicted situations that arise. As with the other emotions, our decisions in this respect are guided by feelings (of care and concern) – which tell us whether things are going well or badly. Another reason why a nurturance drive is not enough is because we do not feel *only* love towards our children, as any parent will attest. The resultant conflicts must be resolved through hybrid cognitive-emotional processes.

Learning how to reconcile the various emotional needs with each other in flexible ways determines the bedrock of mental health and maturity. Consider sustainable romantic partnerships, for example, which require judicious integration of LUST with childlike PANIC/GRIEF-type attachment (think of the Madonna-whore syndrome, which arises from an inability to reconcile sexual and affectionate feelings). Affectionate bonds are also difficult to reconcile with the roving SEEKING system (think of the thrill of novelty), as well as the inevitable frustrations that provoke RAGE (hence the ubiquity of domestic strife), which in turn conflicts with the concerns of nurturant CARE, and so on. Sustaining long-term relationships is just one example of the many challenges that face every human heart. To manage life's problems we use emotions as a compass. It is feeling that guides all learning from experience, in the various forms I have outlined. But biology provides one further drive to help us on our way:

(7) PLAY. We *need* to play. It is the medium through which territories are claimed and defended, social hierarchies are formed, and in-group and out-group boundaries are forged and maintained.

People are often surprised to learn that it is a biological drive, but all juvenile mammals engage in vigorous rough-and-tumble play. If they are deprived of their quota on one day, they will try to make it up the next day – as if by rebound. We all know what rough-and-tumble play is, although the form it takes varies slightly from one mammal species to another. A play session starts with an 'invitation' posture or gesture; then, if accepted, the game is on. The one animal or child exuberantly chases the other, which then stops as they wrestle or tickle each other, taking turns to be on top – accompanied by peals of laughter, or the equivalent vocalisation depending on the species (even rats 'laugh').[28] Then they are back on their feet again, chasing each other in the reverse direction. The associated feeling state is equally universal: it is called fun.

Children love to play. Empirically, however, the majority of play episodes end in tears. This provides an important clue as to what it is all about, biologically speaking; it is about *finding the limits* of what is socially tolerable and permissible. When the game is no longer fun for a playmate, often because they decide you are not being 'fair', they won't play any more. Their limit has then been reached. The marking of such limits is crucial for the formation and maintenance of stable social groups. And the survival of a group is important for the survival of each member of the group, in social species like ours.

A major criterion in this respect is *dominance*. In any play situation, one of the participants takes the lead role and the other is submissive. This is fun for both parties, so long as the dominant one does not insist on calling the shots *all* the time. The acceptable ratio of turn-taking seems to be about 60:40. The '60:40 rule' of *reciprocity* states that the submissive playmate continues playing so long as they are given sufficient opportunities to take the lead.

This reveals a second function of PLAY, namely the establishment of social hierarchies – a 'pecking order'. Rough-and-tumble play accordingly gives way (through development) to more organised and frankly competitive games. Of course, play is not limited to games of the rough-and-tumble variety. We humans engage in pretend play, in which the participants try out different social roles (e.g. Mother/Baby, Teacher/Pupil, Doctor/Patient, Cop/Robber,

Cowboy/Indian,[29] King of the Castle/Dirty Rascal – note the ever-present status and power hierarchies). We do not know what goes on in the imagination of other mammals while playing, but we may confidently hypothesise that they too are 'trying out' different social roles, and thereby learning what they can and can't get away with.

This suggests a third biological function for PLAY. It requires you (and conditions you) to take account of the feelings of others. If you don't, they will refuse to play with you, and then you will be deprived of the enormous pleasure it yields. The bully might get to keep all the toys, but he will be deprived of all the fun. This, it seems, is why PLAY evolved (and why so much pleasure attaches to it): it promotes viable social formations. In a word, it is a major vehicle for developing *empathy*.[30]

Play episodes come to an abrupt end when they lose their 'as if' quality. If you lock up your little sister and throw away the key, then not only have you broken the 60:40 rule but you also are no longer playing the *game* of Cops and Robbers; instead you are imprisoning your sister. In other words, what is governing your mutual behaviour now is FEAR or RAGE rather than PLAY. The same applies to the other games enumerated above. 'Doctor/Patient' is a game until it becomes real sex; then it is governed by LUST. The fact that PLAY hovers, as it were, between all the other instinctual emotions – trying them out and learning their limits – is perhaps the reason why it has not been possible to identify a single brain circuit for PLAY. Probably it recruits them all.[31] Anyone who doubts that playing is definitely a basic instinct, though, should read Sergio and Vivien Pellis's wonderful book, *The Playful Brain*.[32]

We don't always like to recognise that humans, like other mammals, *naturally* claim territories and form social hierarchies with clear rules. (The rules governing primate behaviour are remarkably complex.) The structure of families, clans, armies, even nations – almost any social group – is undeniably hierarchical and territorial; and this has been so throughout history. The higher the social status of an individual within the group, the greater the access that individual has to the resources in the territory the group controls. This observation is not a matter of personal preference; it is a matter of fact. If we do not face such facts, we cannot begin to deal with them. The fact that emotional drives exist does not mean we have no control over them – that we are obliged to bow before 'the Law of the Jungle'; but we ignore these drives at our peril.

It is easy to see how PLAY, in particular, gives rise to social rules. Rules regulate group behaviour, and thereby protect us from the excesses of our individual needs. It is also easy to see how social rules encourage complex forms of communication, and therefore how they contribute to the emergence of symbolic thought. The 'as if' quality of PLAY suggests that it might even be the biological precursor of thinking in general (i.e. of virtual versus real action; see above). Some scientists also believe that *dreaming* is nocturnal PLAY, i.e. trying out the instinctual emotions in an 'as if' world. Interestingly, in REM behaviour disorder, where the motor paralysis which normally accompanies dreams is lost due to brainstem damage, patients (and experimental animals) physically *enact* the various instinctual stereotypes – e.g. fleeing, freezing, predatory pouncing and affective attack.

I hope you can now see that, while bodily affects have a certain vividness and immediacy which makes them seem in a sense unmissable, emotional affects operate through conscious feeling too. Although we do not always recognise them for what they are, they regulate almost all our voluntary behaviour through their various inner spurs and spurts. Voluntary behaviour consists essentially in the making of here-and-now choices. How can you make choices without them being grounded in some evaluative system that tells you which option is better or worse? It is these values that feeling contributes to behaviour.

However, because our background emotional states are not always recognised by the *cognitive* brain, and therefore not always self-reflectively declared, we don't see the larger pattern until we look back and join the dots. When I say 'we' here, I am not referring only to lay people, who have every right to be ignorant of the experimental facts. Although it is true that we all too easily overlook the pivotal role that feelings play in everyday life, I am actually referring to the mainstream of contemporary cognitive science. Cognitive scientists routinely overlook feelings.

But as the next chapter will show from a neurological perspective, any scientific account of consciousness that ignores the fundamental role of feelings misses the main event.

6

The Source

A basic assumption of neuropsychology – grounded in the clinico-anatomical method – is that if a particular mental function is performed by a particular brain region, then a complete lesion of that region must result in complete loss of that function. As I have shown, when it comes to consciousness, the cortex fails this test. But things are even worse for the cortical theory of consciousness: lesions *elsewhere* in the brain obliterate it completely – and very small lesions at that.

The physiologists Giuseppe Moruzzi and Horace Magoun established over seventy years ago that consciousness in cats is lost following tiny incisions that disconnect the cortex from the 'reticulate' (net-like) core of the brainstem.[1] This core must be approximately 525 million years old, because it is shared by all vertebrates – from fishes to humans. Since Moruzzi and Magoun's discovery of it, researchers have confirmed in all manner of species that relatively small lesions in this core – known technically as the reticular activating system – cause coma. For example, David Fischer and colleagues recently identified in human patients with brainstem stroke a tiny, two cubic-millimetre 'coma-specific' region in the upper mesopontine tegmentum (see Figure 1).[2]

There are two possible explanations. The first is that this densely knotted core of the brainstem is where consciousness arises: it is the hidden wellspring of the mind, the source of its essence. I hold this view, as did Jaak Panksepp. The second explanation is that it is like the power cable for a television set: necessary but not sufficient, and hardly enlightening if what you want to understand is how television works. This is the mainstream view.

Suppose the second option is true. By stimulating the brainstem, we might expect to switch consciousness on or off. At best we might attenuate it in certain ways, like a power reduction causing the TV screen to fade. We wouldn't expect it to rewrite the current broadcast on the fly. And yet an electrode implanted in a reticular brainstem nucleus of a sixty-five-year-old woman (for treatment of Parkinson's disease) reliably evoked this remarkable response:

> The patient's face expressed profound sadness within five seconds [...] Although still alert, the patient leaned to the right, started to cry, and verbally communicated feelings of sadness, guilt, uselessness, and hopelessness, such as 'I'm falling down in my head, I no longer wish to live, to see anything, hear anything, feel anything ...' When asked why she was crying and if she felt pain, she responded: 'No, I'm fed up with life, I've had enough ... I don't want to live any more, I'm disgusted with life ... Everything is useless, always feeling worthless, I'm scared in this world.' When asked why she was sad, she replied: 'I'm tired. I want to hide in a corner ... I'm crying over myself, of course ... I'm hopeless, why am I bothering you?' [...] The depression disappeared less than 90 seconds after stimulation was stopped. For the next five minutes the patient was in a slightly hypomanic state, and she laughed and joked with the examiner, playfully pulling his tie. She recalled the entire episode. Stimulation [at another brain site, which was the actual target of the electrode] did not elicit this psychiatric response.[3]

This patient had no previous history of psychiatric symptoms of any kind.

The same applies to *chemical* stimulation or blockade of these core brainstem nuclei. Most antidepressants – serotonin boosters – act on neurons whose cell bodies are located in a region of the reticular activating system called the raphe nuclei (see Figure 1). Serotonin is 'sourced' there, as we say. Antipsychotics – dopamine blockers – act on neurons sourced in another part of the reticular activating system: the ventral tegmental area (see Figure 2). The same applies to anti-anxiety drugs – many of which block a chemical called noradrenaline, which is produced by neurons sourced in the locus coeruleus complex (also Figure 1), yet another part of the reticular activating system. All these neurons are clumped together in the reticulate core of the brainstem. Psychiatrists wouldn't tinker with this region of the brain if it merely switched consciousness on and off. If that were all it did, it would interest only anaesthetists. The second view must therefore be wrong.

Functional neuroimaging of the brain in emotional states points to the same conclusion. Positron emission tomography during states of GRIEF, SEEKING, RAGE and FEAR, for example, show that the highest metabolic activity occurs in the core brainstem (and other subcortical regions; see Figure 9) while the cortex shows *de*activation. Functional magnetic imaging during orgasm reveals the same: the haemodynamic activity that correlates with this intensely affective state is almost exclusively located in the midbrain.[4]

Lesion studies, deep brain stimulation, pharmacological manipulation and functional neuroimaging all point to the same conclusion: the reticulate core of the brainstem generates *affect*. Apparently, therefore, the only part of the brain that we know to be necessary for arousing consciousness as a whole has an equally powerful influence over another mental function, namely feeling. In the previous chapter, I showed how feelings pervade conscious experience: whatever else it is meant to be doing, having and dealing with feelings (which come from *within* us and regulate our biological needs) appears to be one of the central tasks of consciousness. But it now looks as though the neurological

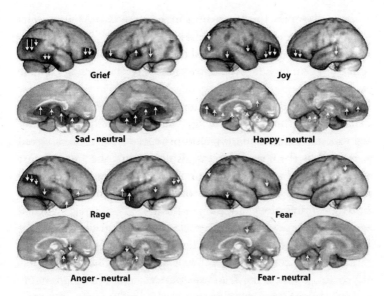

Figure 9 Positron emission tomographic images of four emotional states (courtesy of Antonio Damasio). The upgoing arrows indicate regions of *increased* activation and the downgoing arrows indicate regions of *decreased* activation. The highlighted area in the 'Joy' image appears to show activation of the SEEKING system.

sources of affect and of consciousness are, at a minimum, deeply entangled with one another, and they may in fact be the very same machinery. Contrary to the classical empiricist view, according to which consciousness flows in through our senses, and contrary to the statement I quoted from Meynert that was based on that view, it seems the brain *does* 'radiate its own heat'.

What should we call the basic medium, this mysterious mind-stuff that seems to well up within us? We cannot call it a 'waking state', as Zeman did, since that would require us to describe dreaming as a type of wakefulness, which is absurd. We cannot call it a quantitative 'level' either, as Moruzzi and Magoun did, since the facts just described show that it entails intensely qualitative features.

So, let's try the third term used in the literature: 'arousal'. This seems to me a good, neutral word for it. Both waking and dreaming involve arousal; and it does not preclude quality – as the term 'level' does. In fact, 'arousal' positively suggests feeling.

But what *is* arousal? We have talked about it in straightforwardly behavioural terms – for example, the distinctions between coma, the vegetative state and fully responsive wakefulness. It is usually measured by the Glasgow Coma Scale: tests of the patient's eye-opening responses, their verbal responses to questions and their motor responses to instructions and to pain. However, it can also be defined physiologically.

An EEG produces graphic tracings of cortical electrical activity. When left to its own devices (i.e. if disconnected from the reticular activating system, even when processing sensory inputs), the cortex produces the *delta* wave pattern, a series of high-amplitude waves occurring roughly twice every second (i.e. a frequency of 2Hz). When stimulated by the reticular activating system in the absence of sensory input, the cortex typically produces the *theta* rhythm (4–7Hz)[5] or *alpha* rhythm (desynchronised waves at frequencies of 8–13Hz; 'desynchronised' means erratic). When actively processing external information, the cortex typically displays the *beta* (desynchronised, very low amplitude waves with frequencies of 14–24Hz) or *gamma* pattern (low amplitude waves with very high frequencies of 25–100Hz). Gamma is the rhythm most commonly associated with consciousness.

Nowadays, it is possible also to measure physiological arousal using functional neuroimaging, which literally pictures brain activity by mapping regional patterns of change in metabolic rates. Figure 3 illustrated this technique, with reference to different stages of sleep. The bottom row of the figure shows REM sleep arousal, which is typically (although, as I showed, not exclusively) associated with dream consciousness, and which is sourced in the upper brainstem. Neuroimaging of some of the basic emotional

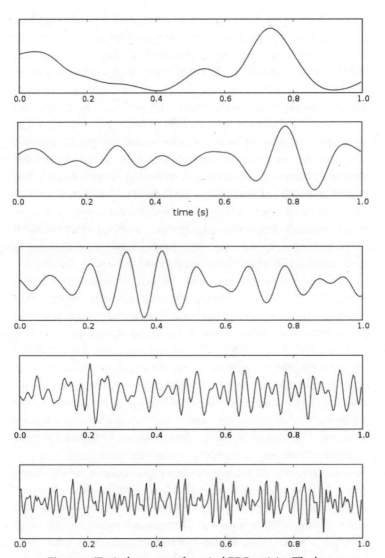

Figure 10 Typical patterns of cortical EEG activity. The least aroused (delta) pattern is shown at the top and the most aroused (gamma) pattern at the bottom. The intermediate patterns, from top to bottom, are theta, alpha and beta.

states of consciousness described above (see Figure 9) and of orgasm, shows the same thing: brainstem arousal.

The cortex becomes conscious only to the extent that it is aroused by the brainstem. The relationship between the two is hierarchical; cortical consciousness *depends* upon brainstem arousal. That is why the delta wave pattern shown at the top of Figure 10 – which is not associated with conscious behaviour – is generated by intrinsic cortical activity, and that is why the gamma rhythm shown at the bottom of Figure 10 – which is strongly associated with consciousness – can be driven by the reticular activating system alone.[6] That is also why the drop in physiological arousal shown in the top row of Figure 3 coincides with waning of consciousness and sleep onset, whereas the increased arousal shown in the bottom row coincides with reappearance of consciousness in dreaming. These facts are not controversial.

Let's peer more deeply into the actual brain mechanisms involved. To do this, I need to introduce a basic distinction between two ways in which neurons communicate with one another. This distinction turns out to be important for consciousness.

Most people with a casual interest in the brain know that neurons transmit messages along intricate networks. This process is called *synaptic transmission*, because it involves passing messages across synapses, the structures by which one neuron conveys signals to another (the word 'synapse' derives from the Greek for 'join together'). Synaptic transmission uses molecules called *neurotransmitters*, which are passed from one neuron to the next, either exciting the post-synaptic neuron or inhibiting it, depending on the molecule in question (glutamate and aspartate are excitatory neurotransmitters and gamma-aminobutyric acid [GABA] is inhibitory). If the downstream neuron is excited by a flurry of neurotransmitters, it passes its own molecules on to the next neurons in the network. If it isn't, it doesn't. Thereafter, the transmitter molecules quickly degrade or are taken back into the presynaptic neuron, to limit the duration of their effect – a process called 'reuptake'.

Synaptic transmission is targeted, binary (yes/no), and rapid. It is the aspect of brain function that is most reminiscent of digital computation, which may explain why it has been such a focus for computationally minded neuroscientists. It occurs all over the nervous system, including the cortex. But it is not intrinsically conscious. In other words, this type of neurotransmission occurs in the cortex whether it is conscious or not. And it has next to nothing to do with arousal.

What fewer people know is that synaptic transmission takes place under the constant influence of a completely different physiological process. This other type of neuronal activity is called *post-synaptic modulation*. Unlike synaptic transmission it is messy, inescapably chemical and very different from what happens in a typical computer. It arises endogenously from the reticular activating system (and other subcortical structures, and even from some non-neurological bodily structures); and it has everything to do with arousal.

The central players in this process are a class of molecules called *neuromodulators*. Unlike neurotransmitters, these molecules spread diffusely through the brain – that is, they are released into the general vicinity of whole populations of neurons rather than at individual synapses.* Instead of passing messages along specific 'channels', they wash over swathes of the network, thereby regulating the overall 'state' of the cortex. For example, the cortex is in a different state in the top and bottom rows of images in Figure 3 (slow wave sleep vs REM sleep) and in the four emotional states shown in Figure 9 (GRIEF vs SEEKING vs RAGE vs FEAR). In each of these states, it processes information differently. Thus, if someone calls your name when you are

---

* Note that some neuromodulatory molecules can act *also* as neurotransmitters.

asleep, you react very differently from how you react when you are fully awake.[7] Likewise, consider your response to an approaching stranger when you are in a state of SEEKING versus FEAR: in the former state you might greet the stranger, and even strike up a conversation with them, whereas in the latter state you might look away and hope they don't notice you.

This distinction between 'channel' and 'state' is a useful short-hand for the two ways in which neurons communicate with each other.[8] The *state* of the cortex affects the differential strengths of the message-passing that goes on within its *channels*; in a manner of speaking, the state adjusts how 'loudly' the different channels communicate (see Figure 11). That is why the same sound (e.g. someone calling your name) is spread widely in the cortex during wakefulness but sequestered in the auditory cortex during sleep, and why a stranger arouses one brain network during the FEAR state and another during SEEKING.

This is the crux of what we call arousal. Note, however, that cortical arousal can be modulated both upwards *and downwards*, to the point of suppressing transmission completely, as happens every night when we go to sleep (which is why some physiologists prefer the term 'modulation' over 'arousal'). Arousal therefore determines which synaptic impulses will be transmitted, and how strongly, as in the example of your name being called when you are asleep versus awake.

Synaptic transmission is binary (on/off, yes/no, 1/0) but post-synaptic neuromodulation *grades* the likelihood that a given set of neurons will fire. It shifts the statistical odds that something will happen in them. This probabilistic, analogue adjustment of firing rates is effected through receptors located at several places along the length of the neuron. Unlike neurotransmitters, neuromodulators have relatively slow-acting and long-lasting effects – not only because the chemicals themselves linger, but because the channels that fire more frequently become more likely to fire over time. If you boost some part of the network, it will stay somewhat

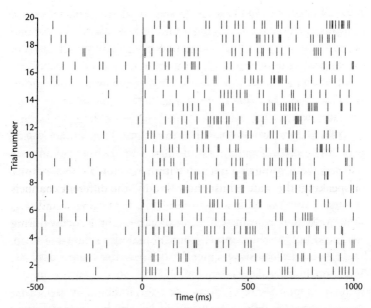

Figure 11 The image above is a plot of the spike trains of twenty
neurons (represented on the vertical *y* axis) over a period of 1.5
seconds (represented on the horizontal *x* axis) during presentation of
a visual stimulus. Stimulus presentation occurs at time 0 on the *x* axis
(represented by the second vertical line). At this point the neurons, which
have a baseline firing rate of 6Hz (on average, when there is no stimulus),
increase their average firing rate to 30Hz. A spike train is a sequence
in which a neuron fires (= spikes) and does not fire (= silences). This
can be represented as a digital sequence of information: '1' for a spike
and '0' for a silence. For example, an encoded spike train could read as
'001111101101'. The first two 0s here represent the latency time between
stimulus presentation and the first spike. The important point to note is
that the firing rates are not determined by the stimulus alone; they arise
from an interaction between the stimulus, the long-term potentiation
or inhibition of the neurons (i.e. whether the stimulus is familiar or
unfamiliar) and their current level of modulation. The adjustment
of post-synaptic modulation is the effect of arousal on neurons. For
this reason, the stimulus depicted here might evoke no response from
the same neurons when the arousal is modulated downwards.

boosted until such time as it is down-modulated. This influences neuroplasticity and it is a big part of how learning works. Arousal states inscribe our various lessons more deeply in the channels of our brains. Thus, for example, you are more likely to remember a journey when you are anxiously trying to find an unfamiliar destination than when you are travelling to the same place by habit, on autopilot.

Where do neuromodulators come from? They come from all over the body, including the pituitary, adrenal, thyroid and sex glands (which produce various hormones) and the hypothalamus (which produces innumerable peptides). But the central source of 'arousal' from the brain's point of view is the reticular activating system. Reticular brainstem arousal releases the five best-known neuromodulators: *dopamine* (sourced mainly in the ventral tegmental area and substantia nigra), *noradrenaline* (sourced mainly in the locus coeruleus complex), *acetylcholine* (sourced mainly in the mesopontine tegmentum and basal forebrain nuclei), *serotonin* (sourced mainly in the raphe nuclei) and *histamine* (sourced mainly in the tuberomammillary hypothalamus). It is no accident that some of these chemicals have featured prominently in previous chapters. Each of the molecules and their associated receptors, of which there are many subtypes, is responsible for a different aspect of modulatory arousal. In addition to these five, there are myriad others – mainly slow-acting hormones and peptides (of which there are over a hundred in the brain), which modulate highly specific neural systems.

The effect of all these modulators is determined by the presence or absence of the relevant receptors and receptor subtypes. In other words, although neuromodulators are spread far and wide through the brain, they only influence cells that possess the relevant receptors. Arousal must not be thought of as a crude or blunt process. It is highly multifaceted, and it unfolds over many dimensions, both spatially and temporally. The potential messages carried by neuromodulators float through the many networks of the brain, but (like painkillers) they are only made use of where they are needed. For example: as you read this, all sorts of background stimuli are being registered by your sensory systems without you paying attention to them, but if you are a mother and your newborn baby starts crying, it overrides your focus on the book and you immediately become aware of the baby. This is due to the increased levels of the specific hormones and peptides – oestrogen, progesterone, prolactin and oxytocin – that float through your brain when you have had a baby, and change its state (i.e. activate the brain's CARE system).

Neuromodulators can only modulate (upwards or downwards) signals that actually exist – i.e. currently active channels. They are released diffusely but they influence only those neurons which (1) have the relevant receptors and (2) are currently active. The same modulator has different effects upon different classes of receptors at different sites. Further specificity is obtained by those modulators that are released not only by the brainstem (and endocrine system) but also in the neural circuits themselves. The five best-known modulators of arousal that I listed above are sourced mainly in the reticular brainstem. The hormones are sourced elsewhere in the body and reach the brain mainly via the bloodstream. The more specific ones are the peptides, many of which are sourced in the hypothalamus. Each peptide does something different, again depending on the receptor type and site. Many of them, which modulate the basic emotions in highly focused ways, are listed in the endnotes to Chapter 5.

So far in this chapter I have addressed two questions: *where* does arousal come from anatomically and *how* does it arise physiologically? The answers are that it is generated mainly, but not exclusively, in the brainstem and hypothalamus, and it arouses the forebrain by modulating neurotransmission. To recap from Chapter 5, *why* arousal is generated is that it responds to endogenous demands made upon the mind to perform work. These demands take the form of a multiplicity of 'error' signals that converge on the core of the brain. Most such signals are dealt with automatically and unconsciously, and demand conscious responses only in situations where automatic responses cannot suffice. This brings me to the main question of the present chapter, namely: where does this seemingly magical shift from automatic reflex to volitional feeling occur?

What I am about to describe and explain is not definitive. That is because the facts themselves are not yet entirely clear. Nevertheless, current neuroscientific knowledge is sufficiently advanced for a broad outline of the overall picture to have emerged. The pioneer of this research was Jaak Panksepp, followed by Antonio Damasio. Damasio has a gift for seeing the big picture, but the neuroscientist to whom I think we owe the greatest debt here, regarding how the brain makes the mysterious leap from autonomic reflex to voluntary action, is Bjorn Merker. He has studied brainstem arousal and orienting mechanisms in a wide range of vertebrate species, including birds, rodents, cats and primates.

The shift from vegetative wakefulness to affective arousal appears to depend upon the integrity of a small, tightly packed knot of neurons surrounding the central canal of the midbrain, – the *periaqueductal grey* (PAG), where *all* the brain's affective circuitry converges (see Figure 6 for the location of this structure). Therefore, patients with localised lesions in the PAG 'stare into empty psycho-affective space':[9]

> Extensive PAG damage [causes] a spectacular deterioration of all conscious activities [...] For instance, early studies in which lesioning electrodes were threaded from the fourth ventricle up the aqueduct to the caudal edge of the diencephalon yielded striking deficits in consciousness in cats and monkeys as operationalized by their *failure to exhibit any apparent intentional behavior* and their *global lack of responsivity to emotional stimuli*.[10] While forms of damage to any other

higher areas of the brain can damage the [cognitive] 'tools of consciousness', they typically do not impair the foundation of intentionality itself. PAG lesions do this with the smallest absolute destruction of brain tissue.[11]

The state just described is the 'vegetative state'. The main difference between it and coma is preservation of wakefulness. But the circadian sleep/wake cycle is just another autonomic function.[12] That is why the vegetative state is also called '*unresponsive* wakefulness' – an apparent oxymoron which reveals the critical distinction between (vegetative) wakefulness and (affective) arousal – what Panksepp called 'intentionality' in the quotation above.[13] This is another reason why I prefer the term 'arousal' over 'wakefulness' or 'level' of consciousness. 'Arousal' accommodates (even positively suggests) emotional responsivity and intentionality, which, as we see here once again, lie at the heart of conscious behaviour.[14] Affective arousal enables *volition*. This is what the PAG adds to automatic, vegetative functioning.

How does it do this? The PAG is not part of the reticular activating system, although it lies right next to it and is densely interconnected with it.[15] The main difference between these nuclei and the PAG is the direction of information flow between them and the forebrain. Whereas the reticular activating system mainly exerts its influence *upwards* into the cortex, the cortex only transmits signals back *down* to the PAG.*

---

* Throughout this book, I am using the words 'above' and 'below', 'higher' and 'lower', 'ascending' and 'descending', etc., not as value judgements but as anatomical locators. Unlike other bodily organs, the brain is *structurally* hierarchical. It is layered somewhat like an archaeological site, with the older levels covered over by the newer ones. Hence the title of this book: The *Hidden* Spring. The deepest core of the brainstem contains the most ancient structures, in evolutionary terms, and the highest levels of the cortex contain the most recent ones. This does not mean that the lower (and older) structures are less important than the higher (newer) ones. On the

The PAG is the final assembly point of all the affect circuits of the brain. So, whereas the forebrain is aroused by the reticular activating system, the PAG is aroused (as it were) by the forebrain. We might think of the reticular activating system and PAG, respectively, as the origin and destination of forebrain arousal.

Accordingly, the PAG is conceptualised as the terminus of the 'descending' network for affect, to be contrasted with the 'ascending' and 'modulatory' affect networks of the brainstem's body-monitoring nuclei and reticular activating system.[16] This means that the PAG is the main *output* centre of all the affective circuits, channelling information to the musculoskeletal and visceral effectors that generate 'emotion proper'.[17] (I am quoting directly from authoritative reviews here, to ensure that I portray our current understanding of these functions accurately.) The descending network for affect is 'involved in specific motor actions invoked in emotions, as well as in the control of heart rate, respiration, vocalisation, and mating behaviour'.[18] The PAG's role in this network is to act as 'an *interface* on salient stimuli between the forebrain and the lower brainstem'.[19] In this respect, the PAG is conceptualised as a centre for '*balancing or segueing* information related to survival salience'.[20] Put differently, the PAG functions 'by *orchestrating* different coping strategies when exposed to external stressors'.[21] It 'provides a massive *assembly point* of the neural systems that generate emotionality'.[22] It thus plays the central role in 'homeostatic defense of the individual's response, *integrating* afferent information from the periphery and information from higher centers'.[23]

Putting it as baldly as I can: *all affective circuits converge on the PAG, which is the main output centre for feelings and emotional behaviours.* That is why 'lower intensities of electrical stimulation of this brain zone will arouse animals to a greater

---

contrary, functionally speaking, the highest forebrain structures are merely *elaborations* of the lowest brainstem ones.

variety of coordinated emotional actions than stimulation at any other brain location'.[24]

To look at it, the poorly differentiated columns that constitute the PAG surround the central canal of the brainstem for a length of 14mm. The central canal, through which a colourless liquid (the cerebrospinal fluid) flows, is the aqueduct of Sylvius – named for the seventeenth-century anatomist who first described it. Its location in the middle of the midbrain gives the PAG its name; 'periaqueductal grey' means simply 'grey matter surrounding the canal'.[25] This primitive core of the brain truly is what Freud described as 'its inmost interior'. It divides into two groups of functional columns.[26] One of them, the back one, is for *active* 'coping strategies' or defensive behaviours such as fight-or-flight reactions, increased blood pressure and non-opioid pain relief.[27] This is where the FEAR, RAGE and PANIC/GRIEF circuits terminate. The front column is for *passive* coping/defensive strategies such as freezing with hyporeactivity, long-term 'sick behaviour', decreased blood pressure and opioid pain relief.[28] The LUST, CARE and SEEKING circuits terminate in this front column.

The PAG is the *final common path* to affective output. It must, therefore, in a word, 'choose' what is to be done next, once the various affective circuits and their associated conditioned behaviours have made their contributions to action.[29] It must make these choices by evaluating the residual error signals that are relayed to it by the affect systems. It must judge their competing bids by the ultimate biological imperatives to survive and reproduce, with each error signal communicating its component need to it. In short, the PAG must *set priorities* for the next action sequence.

But action priorities cannot be determined by needs alone. Needs must be *contextualised* not only by each other but also by the prevailing opportunities.

Take the example of respiratory control, described previously. Consider first the inner context: if I feel suffocation alarm and thirst at the same time, the suffocation alarm must be prioritised over the thirst, but thirst in comparison to sadness, say, might take priority. Secondly consider the outer context: if I experience suffocation alarm, the context will tell me whether I need – for example – to remove an airway obstruction or get out of a carbon-dioxide-filled room. The outside context in which I am acting stays relevant as the events unfold and I decide over and again what to do next.

Remember my definition of affect in Chapter 5; it includes many things that people do not normally think of as affective. Thus, for instance, suffocation alarm and thirst are not merely bodily sensations; they, no less than things like separation anxiety, convey intrinsic values – biological goodness and badness. Pain provides a useful example for distinguishing between the intrinsic valence of affect and its exteroceptive context. The unpleasurable *feeling* of pain is what makes it affective, whereas exteroceptive somatic sensation conveys the *location* of a painful stimulus: 'pain coming from left hand'. This is not a philosophical distinction; the dual aspects can be manipulated separately, for example through stimulation of the PAG and parietal cortex, respectively. It is therefore possible in some clinical conditions to *perceive* that your left hand has been pricked by a pin without *feeling* any pain at all.[30]

Against this background, Bjorn Merker makes a profound point.[31] Although the PAG is anatomically located below the cortex, functionally speaking it is paramount. *After* the cortex and other forebrain structures have performed their cognitive work – after they have submitted their contributions to action – the *final decision* about 'what to do next' is made at the level of the midbrain. Such decisions take the form of affective feelings, generated by the PAG, which can override any cognitive strategies formulated during the previous action sequence. In every moment, the PAG chooses the affect that will determine and modulate the next sequence. For example, on my first day at work as a neuropsychologist, I resolved *cognitively* to return to the beds of the patients and read their files, but what actually happened *affectively* was that I fainted.

Of course, life does not consist in endless emergencies; more often than not, default SEEKING activity (background interest and engagement) will prevail. In emergencies, though, the gaggle of need-conveying error signals arriving at the midbrain cannot all be felt at once, because you cannot *do* everything at once. This, then, is the nub of the PAG's prioritisation function: which of these error signals is the most salient, given current circumstances? Which of my problems can be deferred (or handled automatically) and which must I feel my way through?

The requirement that current circumstances must be taken into account naturally requires some further apparatus. The PAG renders its verdict with the help of an adjacent midbrain structure, known as the *superior colliculi* (see Figure 6). This is located immediately behind it and is divided into several layers, each of which provide simple mappings derived from the body's senses. The deeper layers supply motor maps of the body, while the superficial layers are responsible for spatial-sensory ones. Together they assemble a massively compressed and integrated representation of the exteroceptive world, arriving partly from the cortex but also from subcortical sensory-motor regions such as the optic nerve (again, see Figure 6). The superior colliculi thus represent in distilled form the moment-by-moment state of the *objective* (sensory and motor) body, in much the same way as the PAG monitors its *subjective* (need) state. Merker calls this affective/sensory/motor interface between the PAG, the superior colliculi and the midbrain locomotor region the brain's 'decision triangle'.[32] Panksepp called it the primal SELF, the very source of our sentient being.[33]

Midbrain decisions about what to do next are therefore based upon feedback from the brain's affective circuits *together with* its sensory-motor maps, each of which updates us on different aspects of 'where things stand now'. Recall my example of suddenly becoming aware of my need to urinate at the end of a two-hour lecture. At that point, sensing the opportunity, my bodily need was felt and it became a volitional *drive*. In short, because the midbrain decision triangle takes both internal and external conditions into account, it prioritises behavioural options on the basis not only of current needs but also of current opportunities.

The deepest layer of the superior colliculi consists in a map that controls eye movements. This map is intrinsically more stable than are the overlying sensory maps, for the reason that the other maps are calibrated relative to it, thereby establishing the unified 'point of view' that characterises subjective perceptual experience. That is why we experience ourselves within a stable visual scene no matter how much our eyes dart about, as they do about three times per second. The stabilised scene also hints at the fact that what we perceive is just that – a *scene* – a constructed perspective upon reality, not reality itself. This is also why we experience ourselves as living in our heads.[34] I will explain the virtual nature of perception in more detail in Chapter 10.

As we saw in relation to blindsight, the superior colliculi's 'two-dimensional screen-like map' of the sensory-motor world, as Merker calls it, is unconscious in human beings.[35] It contains little more than a representation of the direction of 'target deviation' – the target being the focus of each action cycle – producing gaze, attention and action orientation. Brian White calls it a 'saliency' or 'priority' map. Panksepp explains that this is how our 'deviations from a resting state come to be represented as states of action readiness'.[36] I cannot put it any better myself.

Perceptual consciousness of the world around us becomes possible with the help of suitably aroused cortex, which (unlike affective consciousness) is what hydranencephalic children and decorticate animals lack. The superior colliculi provide condensed here-and-now mappings of potential targets and actions, but the cortex provides the detailed 'representations' that we use to guide each action sequence as it unfolds. In addition to these highly differentiated images, there are in the subcortical forebrain many *unconscious* action programmes which are called 'procedures' and 'responses' – not images. (Think, for example, of the automatised kinds of memory you rely upon to ride a bike, or to navigate the route to a familiar location.) These are encoded primarily in the *subcortical* basal ganglia, amygdala and cerebellum. Memories are not mere records of the past. Biologically speaking, they are *about* the past but they are *for* the future. They are, all of them, in their essence, *predictions* aimed at meeting our needs. I will take up this important point in the chapters which follow.

The motor tendencies that are activated through midbrain-affect selection release simple reflexes and instincts – and that is all they do in babies, hydranencephalic children and many animals. But, as you now know, such automatic behaviours are brought under individualised control during development, through learning from experience. Stereotyped responses are thereby supplemented by a more flexible repertoire of options. The behavioural sequence arising with each new action cycle unfolds upwards over these progressively expanding levels of forebrain control, from procedural 'responses' to representational 'memory images'. This generates what Merker calls a 'fully articulated, panoramic, three-dimensional world composed of shaped solid objects: the world of our familiar phenomenal experience'.[37]

Recall that long-term memories serve the future. Once the midbrain decision triangle has evaluated the compressed feedback flowing in from each previous action, what it activates is an expanded *feedforward* process which unfolds in the reverse direction, through the forebrain's memory systems, generating an *expected context* for the selected motor sequence. This is the product of all our learning. In other words, when a need propels us into the world, *we do not discover that world afresh with each new cycle*. It activates a set of predictions about the likely sensory consequences of our actions, based upon our past experience of how to meet the selected need in the prevailing circumstances.

Voluntary action then entails a process of testing our *expectations* against the *actual* consequences of our actions. The comparison produces an error signal, which we use to reassess our expectations as we go along and adjust our action plans accordingly. This is what 'voluntary' behaviour is all about: deciding what to do in conditions of uncertainty, mediated by the consequences of each action. This involves both affectively felt and perceptually sensed consequences, which is why both affective and sensory-motor residual error signals converge on the midbrain decision triangle. I like Jakob Hohwy's term for the mental process that controls voluntary behaviour: 'predicting the present'.[38]

To get a clinical impression of what I am talking about, consider what happened with Mr S, the electrical engineer with Korsakoff psychosis whose case I described in Chapter 2. He expected to see Johannesburg when he looked through my window. This was due to his memory disorder: he had not updated his predictive model in line with recent events. When the snowy London landscape that he actually saw failed to match his expectations, he didn't change his mind. He ignored the incoming error signal and stuck with his original prediction, saying 'No, I *know* I'm in Jo'burg; just because you're eating pizza, it doesn't mean you're in Italy.' In other words, his brain down-modulated the error signal. He accordingly didn't adjust his action programme, which was bound to end in tears – because he failed to learn from experience. If it were not for the fact that his family and doctors looked after his every need, he would surely have died.[39]

Most people don't realise that our here-and-now perceptions are constantly guided by predictions, generated mainly from long-term memory. But they are. That is why far fewer neurons propagate signals from the external sense organs to the internal memory systems than the other way round.[40] For example, the ratio of incoming connections to outgoing ones in the lateral geniculate body (which relays information from the eyes to the visual cortex and vice-versa; see Figure 6) is about 1:10. The heavy lifting is done by the predictive signals that *meet* the sensory ones arriving from the periphery. This saves an enormous amount of information processing and therefore metabolic work. Considering that the brain consumes about 20 per cent of our total energy supplies, this is a valuable efficiency. Why treat everything in the world as if you'd never encountered it before? Instead, what the brain does is propagate inwards only that portion of the incoming information which does *not* match its expectations. That is why perception is nowadays sometimes described as 'fantasy' and 'controlled hallucination'; it begins with an *expected* scenario which is then adjusted to match the incoming signal.[41] In this sense, the classical anatomists were right: cortical processing consists mainly in the activation of 'memory images', suitably rearranged to predict the next cycle of perception and action.

What I am saying here applies to all vertebrates, not only to humans. It might even apply to all organisms equipped with a brain, or a nervous system. (Insects, for example, have brain structures that function much like our reticular activating system.)[42] We know by now that the fundamental form of consciousness is affect, which enables us to 'feel' our way through unpredicted situations. But how does endogenous affect turn into conscious exteroception?

Panksepp suggests that an evolutionary bridge between these two aspects of experience could have been forged by the 'sensory affects' (e.g. pain, disgust and surprise). Sensory affects are simultaneously internal feelings *and* external perceptions; they are inherently valenced perceptions qualified by specific feelings. So, for example, pain feels different from disgust, and you respond to them differently, withdrawing or retching, depending on which one you have. Over this evolutionary bridge, in Panksepp's view, consciousness became extended onto perception in general, since it contextualises affect. After all, the external world acquires value for us only because we must meet our needs there; it acquires value *in relation to* affect.

So, my answer to the question of why perception, action and cognition are ever felt at all is that *they are felt because they contextualise affect*. It's as if our perceptual experience says: 'I feel like this *about that*.'[43] Perception is, as it were, applied uncertainty. This makes it reasonable to say, as Merker does, that affective and perceptual consciousness utilise a 'common currency'; they are different *kinds* of feeling, but they are feelings nonetheless.[44] I am talking only about *conscious* perception here. It is a small step from there to infer that our five modalities of perceptual consciousness (sight, hearing, touch, taste and smell)[45] evolved to qualify the different categories of external information that are registered by our sensory organs,[46] just as the seven varieties of affective consciousness discussed in the previous chapter qualify the different categories of emotional need in mammals.

But it is important to recall also that cortical perception is *apperception*. What appear in consciousness are not the raw sensory signals transmitted from the periphery, but rather predictive *inferences* derived from memory traces of such signals and their consequences. The fact that feeling is spread, as it were, over stabilised cortical inferences to create what I call 'mental solids'[47] – the external world as it manifests in perception – partly accounts for the different phenomenal qualities of affect versus conscious perception. I will pick up these themes again in Chapter 10.

The brain's internal model is the map we use to navigate the world – indeed to *generate* an expected world. But we can't take all our predictions at face value. There are in fact two aspects to the 'expected context' that the internal model generates: on the one hand we have the actual content of our predictions, and on the other is our *level of confidence* about their accuracy. Since all predictions are probabilistic, the degree of *expected uncertainty* attaching to them must also be coded. The predictions themselves are furnished by the forebrain's long-term memory networks, which filter the present through the lens of the past. But the second dimension – the adjustment of confidence levels – is the essence of the work that is performed by modulatory arousal.

This is what went awry in the case of Mr S: he placed too much confidence in his predictions. To say the same thing differently, as a result of his condition, he gave insufficient weight to his error signals (he paid too little *attention* to them).[48] How can that be? In a sense we have already looked at this from the other side. Modulating neuronal signals just means adjusting their strength up or down. That's what the midbrain decision triangle does when it selects which signals to boost; it *modulates* signal strength via the reticular activating system. Baseline signal strengths represent the *expected* context. However, as the actual (experienced) context unfolds, the signal strengths must be adjusted. It follows that increased confidence in an error signal necessarily implies decreased confidence in the prediction which led to the error. If your smoke alarm goes off but nothing appears to be burning, either the alarm or the appearances must be at fault, and somehow a verdict must be rendered on the disagreement. That is what happens in the midbrain decision triangle: the competing claims are assessed and a winner is declared. The outcome of this process is conveyed to the forebrain by the reticular activating system, which acts on the basis of your expectations and then – as the selected action unfolds – releases clouds of neuromodulating molecules to fine-tune the signals in your long-term memory networks, up-regulating the forebrain channels in which some predictions are stored and down-regulating others. This, in turn, leads to learning from experience. In this way, we keep improving our generative model of the world, by trial and error.

Everything we do in the realm of uncertainty is guided by these fluctuating confidence levels. We think we know what will happen if we act in a certain way, but do we really? If our conviction drops below a certain threshold, we won't act, or we'll change tack. In the exteroceptive sphere, things are going well when they turn out as expected, and badly when uncertainty prevails. Accordingly, things *feel* good or bad: increasing confidence (in a prediction) is good, decreasing confidence is bad. So, we try to minimise uncertainty in our expectations.

As we saw above, perception and action are an ongoing process of *hypothesis-testing*, in which the brain constantly tries to suppress error signals and confirm its hypotheses. (Experimental science itself is just a systematised version of this everyday process, as we saw in Chapter 1: 'if hypothesis *X* is correct, then *Y* should happen when I do *Z*'.) The more your hypotheses are confirmed, the more confident you are, and the less aroused – less conscious – you need to be. You can automatise your action sequences and drift off into the default mode. But if you find yourself in an unexpected situation – one in which your predictive model appears to shed no reliable light – the consequences of your actions become highly salient. You switch out of autopilot and become hyper-aware: the decision triangle carefully adjusts your predictions as you feel your way through the consequences of your actions and make new choices.

Feeling thus remains the common factor in all consciousness, both affective and cognitive. Its function is to evaluate the success or failure of your action programmes and their associated contexts. But not all failures are equally significant. The strongest feelings relate to uncertainties of survival value rather than those attaching to relatively minor eventualities. Bad calls such as 'the car I am following will turn left now' may be dealt with at peripheral levels in the predictive hierarchy, where greater error is tolerated, and thereby entail shifts of what we call attention rather than affect (although attention is equally modulated by the reticular brainstem). If the car you are following suddenly seems likely to involve you in an accident, however, your error is more likely to arouse an affective response. Attention can be both directed and grabbed. In the latter case it is accompanied by feelings of startlement or fright, which rewrite the failed predictions on the fly, to make sure that your model of the world doesn't spring the same nasty surprise on you again.

That, on a physiological level, is how consciousness works in the brain. It might not seem altogether satisfying. For one thing, I haven't bothered too much with philosophical approaches to the mind/body problem. I haven't addressed the relationship between what some philosophers call its 'dual aspects': why is the objective physiology of consciousness accompanied by a subjective phenomenal feel? When scientists observe a regular correlation like this, between any two data sets, they look for a single underlying cause. They want to *explain why* the phenomena co-occur. To understand the regular conjunctions between the physiological and psychological aspects of consciousness, therefore, we need to go deeper. We must go beyond the disciplinary constraints of psychology and physiology. We scientists typically do so by turning not to metaphysics but to physics.[49] In this case, the answers we require are to be found in the physics of entropy.*

---

* Readers who have not been following the endnotes might want to look at the Appendix on p. 306 at this point. Although somewhat technical, it provides a useful bridge to the next chapter.

7

The Free Energy Principle

A group of scientists are in a laboratory, watching a large computer monitor. Dots and spots are swirling about on the screen. You can make out different colours: blue, red, purple and more. The dots seem to come in different sizes too, but you cannot discern any pattern in their swarming motions. They billow and interpenetrate like clouds of gas, haphazardly filling the virtual space. Nevertheless, a digital clock marks the passage of time in seconds, and two axes along the bottom and left side of the screen identify the spatial location of each dot. This suggests that something measurable is going on. But how can you quantify the movements in this chaos?

You ask one of the physicists in the room what you are looking at and she says, unhelpfully, that it is a 'stochastic' process. Stochastic means random. Gradually you notice that the smaller dots appear to be more sluggish than the larger ones; they seem to be dancing to a slightly different tune. On the instructions of a neuroscientist in the room, the computer technician enters something on the keyboard and the swirling screen comes to a sudden halt. The technician carefully saves the data. Then, again at the neuroscientist's behest, he keys in a flurry of numbers which change the values on a set of equations that now appears on a small screen alongside the larger one you were watching. He explains that he is adjusting the 'local interactions between the subsystems'.

Once again the cloud-like particles enter the bottom-left corner of the screen and start swirling. This time, after a brief period of chaos, it is easier to discern a pattern. The coloured dots billow mainly outwards at first; then, gradually, they coalesce into clumps, and spontaneously converge upon the centre of the screen to form a blob-like mass. With a little imagination, it could be a compact flock of birds (starlings perhaps) wheeling in formation. Steadily, the movement of the particles becomes more constrained. They jostle with each other like soldiers on a parade ground taking up assigned positions, until a clear structure begins to emerge: four concentric layers. At the centre are dark-blue dots. These are surrounded by red ones, which in turn are surrounded by purple ones. Light-blue ones form a sort of outer boundary. All of this happens against a background of smaller black dots that seem to continue drifting about aimlessly. The neuroscientist looks pleased. He asks the technician to freeze the screen. The clock says that 1,278 seconds have passed.

This is the Wellcome Centre for Human Neuroimaging at Queen Square in London, where Karl Friston is the scientific director. He has kindly invited you to observe this interesting experiment: a simulation of the short-range interactions that occur between different physical subsystems when they are subjected to various forces.

The rules that govern these virtual particles are of the same broad character as those governing the behaviour of real atoms and molecules: mindless (but not indiscriminate) propensities to attract and repel one another. Such interactions evidently produce *order from chaos*. Spontaneous ordering of this kind is thought to have occurred when life emerged from the primal soup. Yet the experiment is being conducted in a centre for cognitive neuroscience, across the lawn from the National Hospital for Neurology and Neurosurgery, and you could be forgiven for wondering what these virtual particles have to do with the brain.

Friston explains that biological systems such as cells must have emerged through complex versions of the same process that formed simpler 'self-organising' systems such as crystals from liquid, because they share a common mechanism. This mechanism, he says, is 'free energy minimisation' (which I will explain shortly). All self-organising systems, including you and me, have one fundamental task in common: to keep existing. Friston believes that we achieve this task by minimising our free energy. Crystals, cells and brains, he says, are just increasingly complex manifestations of this basic self-preserving mechanism.[1] Indeed, so many aspects of what we think of as mental life appear at the very dawn of biological organisation that the contribution made by actual brains can start to seem rather subtle. But, so long as we keep a tight hold of the concept of free energy, all – really, all – will become clear.

Karl Friston is the last of the great scientists who have shaped my life's work. He is in my opinion a genius and (objectively) the most influential neuroscientist in the world today. Influence is measured by your 'h-index', which measures the impact of your publications.[2] As a rule of thumb, when your h-index is greater than the number of years since you obtained your doctorate, you're doing well. Friston's is 235, the highest of any neuroscientist.[3] His original claim to fame was 'statistical parametric mapping', which enabled the analysis of functional neuroimaging so prevalent today. However, his work on 'predictive coding' and the Free Energy Principle brought him much more renown.

Despite Friston's towering reputation, for many years I took only a distant interest in his work. Then, in 2010, he published a paper with a young psychopharmacologist named Robin Carhart-Harris, whom I knew slightly from his interest in neuropsychoanalysis. His paper with Friston argued that Freud's conception of drive energy (i.e. 'psychical energy') was consistent with the Free Energy Principle.[4] As I explained before, Freud readily admitted that he was 'totally unable to form a conception' of how bodily needs could become a mental energy. He also wrote that this energy was capable of increase, diminution, displacement and discharge, and therefore possessed all the characteristics of a quantity, 'though we have no means of measuring it'.[5] Considering that Freud's original (abandoned) intention had been to 'represent psychical processes as quantitatively determinate states', Carhart-Harris and Friston's paper came as a thunderbolt for me. If mental energy really was isomorphic with changes in thermodynamic free energy, then, I thought, it must be both measurable and reducible to physical laws.

I therefore immersed myself in Friston's earlier publications and sought him out. We met several times over the subsequent years, in London and Frankfurt. The main topic of our conversations was the role of affect in mental life. Because Friston's work at that time was, like nearly everyone's, still heavily corticocentric, the predictive mechanisms he had discovered concerned *cognition* almost exclusively. That is why, for example, a celebrated paper of his showing that predictive coding explains the way in which neurons communicate with each other was entitled 'A theory of *cortical* responses'.[6] I must confess, though, I had never taken the time to fully digest many of his more technical publications.

In 2017 Friston was invited to be the keynote speaker at our annual Neuropsychoanalysis Congress (which was held that year at the old University College Hospital, London, and focused on the topic of predictive coding). If I was going to make my usual Closing Remarks without embarrassing myself, I had to master the physics. Among many of Friston's other publications, therefore, I carefully reread a highly technical article of his in one of the journals of the Royal Society. It was entitled 'Life as we know it'.[7] With great effort, I properly understood it for the first time. This article aimed at nothing less than reducing to mathematical equations the basic laws governing *intentionality*.

The implications were electrifying. It seemed to me that these equations might provide the breakthrough I was looking for. Immediately after our scientific exchanges at the 2017 congress, therefore, I wrote to him suggesting that we pool our insights and attempt to incorporate consciousness under the Free Energy Principle. To my delight, Friston agreed, and we began collaborating on a paper setting out what became our shared view.[8]

The link between Friston's work and my own is *homeostasis*. I explained earlier that we must remain within our physiologically viable bounds. Let's use thermoregulation as a model example. You can't settle for any old body temperature: you must remain within the limited range of 36.5–37.5°C. If you get much hotter than that, you die; if you get much colder, you die. You cannot allow your core body temperature to equalise with the ambient temperature, as hot water does when you add it to a cold bath. Hot water entering the bath doesn't remain separate from the cold in a large globule under the tap. But *you* do – you must, to stay alive – and this requires *work*. Comatose patients cannot perform such work, and so they die from conditions like hyperthermia; they literally overheat.[9] The same applies to the regulation of blood gases, fluid and energy balance, and many other bodily processes. It applies even to *emotional* needs, which, as we saw in Chapter 5, are no less 'biological' than bodily ones. Remaining within the viable bounds of our emotions also requires us to work: to maintain close proximity with our caregivers, to escape from predators, to get rid of frustrating obstacles and so on. Beyond a certain level of predictability, the work required to do these things is regulated by feelings.

The mechanism I have just described is an extended form of homeostasis, and it isn't complicated (see Figure 12).

Every homeostat consists in just three components: a *receptor* (which measures temperature, in my model example), a *control centre* (which determines how to maintain temperature within the viable bounds: 36.5–37.5°C, in my example) and an *effector* (which performs the work required to return you to those bounds when you exceed them). Because the mechanism of homeostasis is so simple, it can be reduced to physical laws. That is what

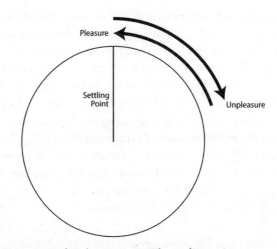

Figure 12 Feeling homeostasis. The settling point
represents the system's viable bounds.

Friston's paper was about: the basic laws governing 'life as we
know it'.

What excited me was the realisation that these laws – suitably
extended to accommodate the less predictable form of homeostasis that underlies affect – might also explain consciousness: and
not only the external observables of conscious behaviour (the
objective *physiology* of the midbrain decision triangle and reticular activating system), but also the internally observed ones: it
promised to explain the subjective *feelings* that guide decisions.

This is where Damasio parted company with me. Perhaps
more than any other neuroscientist, he had drawn attention to
the fact that affect (and therefore consciousness) is, at bottom, a
form of homeostasis. But he rejected the further inference I drew:
if consciousness is homeostatic, and homeostasis is reducible to
physical laws, then so are the phenomena of consciousness.[10] The
mechanism of consciousness, no less than that of the movement
of heavenly bodies and everything else in nature, *must* be lawfully

explicable and therefore somehow predictable, even if only prob-abilistically. When Damasio read a draft of the article that I wrote with Friston, we had an anguished conversation in his office. He couldn't understand why I was trying to reduce consciousness to what he called 'algorithms'. This was a classic case of a scientist baulking at the implications of an insight that he himself visited upon the world – like Einstein's response to quantum mechanics: 'God does not play dice with the universe.'[11]

I briefly put to Damasio the argument that I am going to explain to you now.[12] To do so properly, however, I must first intro-duce you to the physics I had to master to understand the Free Energy Principle. Take a deep breath.

The essence of homeostasis is that living organisms must occupy a limited range of physical states: their viable states, or valued or preferred states, or what Friston calls (referring to all of the above) their 'expected' states. We cannot afford to disperse our-selves across all possible states. It turns out that this biological imperative has a deep link with one of the most basic explana-tory concepts in physics, namely *entropy*. Most people have an intuitive understanding of what 'entropy' is. They think of it as a natural tendency towards disorder, dissipation, dissolution, and the like. The laws of entropy are what make ice melt, batteries lose their charge, billiard balls come to a halt and hot water merge with cold.

Homeostasis runs in the opposite direction. It *resists* entropy. It ensures that you occupy a limited range of states. That is how it maintains your required temperature, and how it keeps you alive – how it prevents you from dissipating. Living things *must* resist one of the fundamental principles of physics: the Second Law of Thermodynamics.

The First Law of Thermodynamics concerns the conservation of energy.[13] It states that energy cannot be created or destroyed; it can only be converted from one kind to another and flow from one place to another. (We also know, thanks to Einstein, that it can be converted into matter.)

The Second Law stipulates that natural processes are always irreversible.[14] Thus the hot bathwater mixing with the cold cannot become re-separated. Likewise, the energy in heat cannot be put back into the burned coal that produced it, and the energy that was wasted in the process cannot be returned to it. This is due to entropy, which accordingly *always increases* on the large scale.[15] Entropy may in fact be the physical basis for the fact that time itself appears to have a direction and a flow.

In thermodynamics, there are two conditions of energy: useful and useless. The 'usefulness' of energy is defined by its capacity to perform work. For example, the energy in a lump of coal can be burned to produce heat,[16] which can boil water to produce steam, which can drive an engine; but at each step of this process, some energy will be lost. That is, you can never usefully employ all of it in work.

Combining these facts, we learn the following: as the useful energy in a system runs down, its entropy increases. This means that the capacity of the system to perform work always decreases. Entropy is therefore associated with loss of useful energy, because the energy is no longer available to perform work. The Second Law is a statement of the ineluctable fact that some energy will be lost to useful work during any natural process.[17]

I mentioned some intuitive understandings of 'entropy' a moment ago, but the formal, technical way in which entropy is defined in physics concerns the number of distinct states that a system can occupy.[18] The entropy is determined by the number of possible microscopic states that would give rise to the same macroscopic state. Simply put: *the fewer the possible states, the lower the entropy.*

Homeostasis sets bounds on the range of macroscopic states that a system such as you and I can occupy. Recall that homeostasis keeps us alive by performing effective work; therefore, if entropy entails the loss of the capacity for work, it is 'bad' from the viewpoint of us biological systems. The most basic function of living things is to resist entropy.

The example that is conventionally used by physicists to illustrate these concepts involves a compressed gas being introduced into an empty, larger chamber through a small aperture. As the molecules move randomly, they explore the chamber, spreading out to occupy the available space. The more time passes, the more locations there are in which each molecule might be found. The only way this process can be reversed – the only way to gather the gas back into its original container – is with work.

Think of entropy in terms of the number of locations that each molecule might be in at a given time. This turns out to be a statement about *probability*. The statistical chance that each molecule will occupy a specific position decreases as entropy increases: as the gas expands, the position of each molecule becomes less predictable. Increasing entropy means *decreasing predictability*.

This is important, because, unlike the other laws of thermodynamics, the laws of probability apply to all things, not just material things. Just as the entropy of a gas in a chamber can be defined probabilistically, so too can the entropy associated with a psychological decision-making process. In both cases the entropy increases with the randomness of possible outcomes. The 'entropy' associated with expanding gases and expanding options *is the same thing*. Not all things that exist in nature are visible and tangible, but they submit to the laws of probability all the same. This is why probability cuts to the heart of modern physics, where matter is no longer considered a fundamental concept and classical particles have disappeared.[19]

In my physical example of entropy, when a gas was initially compressed into its container and its molecules were packed close together, fewer *bits of information* were required to describe the actual location of each molecule than were required after it was released to fill all the available space in the large chamber. In information science, a 'binary digit' – a term that conveniently contracts to 'bit' – is the basic unit of information. A bit can take one of two opposing values, e.g. yes vs no, on vs off, positive vs negative. These states are usually represented as 1 vs 0.[20]

Fewer bits of information are required to describe the gas initially, but as it expands, the number of possible states that each molecule can occupy increases. Thus, entropy (which is usually measured physically in terms of heat change together with temperature) can also be measured in bits: the more information required to describe the microstate of a system (i.e. the state of each and every molecule), the greater the thermodynamic entropy. Or to put it more simply still: *the more yes/no questions one needs to answer to describe a system, the greater its entropy*. Thus, when the microstates of a system have low probability values, each measurement of the system carries more information than it would if they had high probability – because more binary questions would have to be answered to describe the totality of the system's state.

Entropy is minimal when the answer to every yes/no question is entirely predictable, i.e. when nothing is learnt and there is no information gained. The information content of a fair coin toss is one bit, because the chances of the coin coming up heads are 50 per cent, whereas you get zero information tossing a coin with two heads, because the chances of it coming up heads are 100 per cent. It provides no information because the answer is entirely predictable. Entropy measures the average amount of information you get upon multiple measurements of a system. Thus, the entropy of a series of measurements is its *average information*, its average *uncertainty*.

For a glimpse of the importance of this for a neuroscientific understanding of consciousness, remember that synchronised slow-wave patterns on an EEG are more orderly (more predictable) than desynchronised (erratic) fast ones. The 'low-arousal' patterns therefore carry less information than the 'high-arousal' ones (see Figure 10). *The high-arousal ones contain more uncertainty.*[21] Thus EEG entropy values are higher in minimally conscious than in vegetative patients.[22] That makes sense: cortical activity in the conscious brain communicates more information than it does during deep sleep. But here comes the strange part: if more information means more uncertainty and therefore more entropy, then – since living things must resist entropy – waking activity is less desirable, biologically speaking, than deep sleep.[23] I know this is counter-intuitive, but it will become more comprehensible as we proceed.[24]

The relationship between entropy and information was formalised in a famous equation by the electrical engineer and mathematician Claude Shannon. With this breakthrough, Shannon single-handedly incorporated 'information' into physics, where it has since become a basic concept, especially in quantum mechanics.[25] On the basis of Shannon's work, the physicist Edwin Thompson Jaynes argued that thermodynamic entropy should be seen as an *application* of information entropy.[26] Shannon's definition is therefore more fundamental than the thermodynamic one: an abstract definition of entropy in terms of information dynamics is more generally applicable than a concrete one in terms of heat dynamics. The laws of thermodynamics could therefore be seen as *a special case of the deeper laws of probability*. This is important because thermodynamic laws apply only to material (tangible, visible) systems, such as brains, while information laws apply also to immaterial (intangible, invisible) systems, such as minds.

But probability is not quite the same as information. Information, in Shannon's sense, entails the additional factor of *communication*. Hence the title of his seminal article that established information science: 'A mathematical theory of communication'.[27] Unlike probabilities – which exist in and of themselves – communication requires both an information *source* and an information *receiver*. (The communicator doesn't have to be a person. It could be a book, for instance, or any system possessing information that a receiver can learn from.)

This poses big problems for any theoretical assumption that consciousness *just is* information.[28] It raises the question: what is the source and what is the receiver of the information – integrated or otherwise? That is why I am dissatisfied with the information-flow models used by cognitive scientists. They elide the question: where is the *subject*, the receiver? In this way, to paraphrase Oliver Sacks, the psyche is excluded from cognitive science.

This omission of the experiencing subject raises a bigger question than that, however – perhaps the biggest one faced by cognitive science today: without an observer, how and why does information processing (i.e. question asking and answering) come about in the first place?

Shannon's discovery of information-as-entropy led the physicist John Wheeler to propose a 'participatory' interpretation of the universe.[29] According to Wheeler, things only come into being in the form they do (i.e. as observable phenomena) in response to the questions we ask. Phenomena, as such, exist only in the eye of a beholder, a participant observer, a question-asker. To use Wheeler's famous phrase: 'Its arise from bits' (where 'its' are observable *things*).[30] Information is therefore 'physical' not only because it is involved in the laws of physics, but also because it is the basis of all *observable phenomena*. That is how abstract forces and energies become observable and measurable: 'That which we call reality arises in the last analysis from the posing of yes/no questions and the registering of equipment-evoked responses; in short […] all things physical are information-theoretic in origin.'[31]

The sensory modalities of the nervous system generate 'equipment-evoked responses' to the questions we ask of the universe. The sensory responses give rise to the phenomena – the 'things' – we experience. Experience itself therefore arises from communication between an information receiver (a participant observer) and an information source; between a questioner and the answers they register. But this still leaves the question: *where do questioners come from?*

These are heady issues. Before I go on, let's pause to take stock. I have conveyed three important points. The first is that the average information of a system is the entropy of that system (i.e. the entropy in a system is a measure of the amount of information needed to describe its physical state). The second is that living systems must resist entropy. These two facts together imply that *we must minimise the information that we process*. (Here I mean information in Shannon's sense, of course; in other words, we must minimise our *uncertainty*.) Everything else I am going to say in this chapter, and the next two, follows from this simple but startling conclusion.

This leads to the third important thing we have learnt so far: we living systems resist entropy through the mechanism of homeostasis. In short, we receive information about our likely survival by *asking questions* (i.e. taking measurements) of our biological state in relation to unfolding events. The more uncertain the answers are (i.e. the more information they contain) the worse for us; it means we are failing in our homeostatic obligation to occupy limited states (our expected states).

The nature of the questions we ask is determined in part by our *species*. Sharks can respire under water but humans can't. So, we have different needs, and we expect to occupy different states. Such requirements are determined by natural selection. Staying within one's evolutionary niche, broadly conceived, is what homeostasis is all about. Which is why each species needs to ask questions such as, 'Am I able to breathe here?' Our very survival depends upon the answers we receive.

Incidentally, why should we think of biological requirements as *expectations*? This form of language may seem odd at this point, but it indicates a deep continuity that will prove important later. If it helps, try taking the perspective of evolution itself, rather than that of an individual creature. Natural selection fitted each species to its ecological niche: each creature's survival depends only on things that are in fact reliably found in its natural habitat. So, we need air *because* we can expect it.

Now I can return to the profound question raised above: where do participant observers come from? In other words: how and why, in physical terms, does question-asking arise?[32]

Here is a very brief history of the idea of self-organisation. The first person to use the term was Immanuel Kant, in his 1790 *Critique of Judgment*. Kant was arguing that living beings have intrinsic 'aims' and 'purposes', which could, he said, only be true if their constituent mechanisms were simultaneously both ends and means. Such 'teleological' entities (i.e. entities with intrinsic aims and purposes), Kant said, must behave intentionally: 'Only under these conditions and upon these terms can such a product be an organised and self-organised being, and, as such, be called a physical end.' How such beings might arise, he believed, was beyond the power of science to explain: there could never be 'a Newton of the blade of grass'.

Then Darwin discovered natural selection. As we now know, natural selection gives rise to the intrinsic aims and purposes of *survival* and *reproduction*. Both of these things turn out to be manifestations of self-organisation.[33] With Darwin's insight, the question of the origin and composition of teleological beings became tractable to science.[34] All that remained was to flesh out the details.

An important further step came in the mid-twentieth century, when Norbert Wiener, the mathematician who founded the discipline of 'cybernetics', added the notion of *feedback* to Shannon's understanding of information. According to Wiener, a system could attain its goal (its 'reference state') by receiving feedback about the consequences of its actions. The feedback includes error signals – measuring *deviations* from the reference state – which would be used to adjust the system's actions, and keep it on course. Homeostasis thus turns out to be a specific case of a more general cybernetic principle: it is a kind of negative feedback.

William Ross Ashby used this notion of feedback, combined with the statistical physics introduced above, to reveal how self-organisation develops naturally.[35] Ashby showed that many complex dynamical systems automatically evolve towards a *settling point*, which he described as an 'attractor' in a 'basin' of surrounding states. The further evolution of such systems then tends to *occupy limited states*.

Hopefully, this tendency to occupy limited states sounds familiar to you: it is nothing other than a tendency to resist entropy. According to Friston, it is this tendency that triggers ever more elaborate forms of self-organisation. Having created the possibility for it between the subsystems in his simulated primal soup, described at the beginning of this chapter, he observed their behaviour developing in three stages:

(1) with some short-range parameters the dots just darted all over the place;
(2) with other parameters they coalesced into stable crystal-like structures;
(3) with yet others they showed more complex behaviours: after coalescing, they jostled restlessly with each other, taking up specific positions within a dynamic structure.

Here is how Friston described it in his own words (don't worry about the technical language; I just want to give you a visual impression of what he saw):

These behaviours range from gas-like behaviour (where subsystems occasionally get close enough to interact) to a cauldron of activity, when subsystems are forced together at the bottom of the potential well. In this regime, subsystems get sufficiently close for the inverse square law to blow them apart – reminiscent of subatomic particle collisions in nuclear physics. With particular parameter values, these sporadic and critical events can render the dynamics non-ergodic, with unpredictable high amplitude fluctuations that do not settle down. In other regimes, a more crystalline structure emerges with muted interactions and low structural (configurational) entropy. However, for most values of the parameters, ergodic behaviour emerges as the ensemble approaches its random global attractor (usually after about 1000 s): generally, subsystems repel each other initially (much like illustrations of the big bang) and then fall back towards the centre, finding each other as they coalesce. Local interactions then mediate a reorganisation, in which subsystems are passed around (sometimes to the periphery) until neighbours gently jostle with each other. In terms of the dynamics, transient synchronisation can be seen as waves of dynamical bursting [...] In brief, the motion and electrochemical dynamics look very much like a restless soup (not unlike solar flares on the surface of the sun) – but does it have any self-organisation beyond this?[36]

The answer to his last questions proves to be yes. A complex dynamical structure emerges, in which dense subsystems, after separating themselves off from their surrounding milieu, form *concentrically layered structures*, each with an inner core and an outer surface, which is further partitioned into two sublayers (see Figure 13). The sublayers of the partitioned surface display highly interesting patterns of interaction with the inner core and with the surrounding milieu, respectively. The states of the outer sublayer are influenced by those of the external milieu and they in turn influence those of the internal subsystems, but this influence is not reciprocated (in other words, the internal constituents of

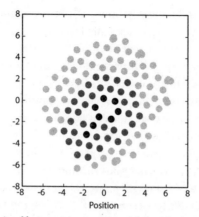

Figure 13 A self-organising system with its Markov blanket. In this image, the inner core of the system is depicted by the black dots and its surrounding layers are depicted by the dark grey ones: the blanket. The light grey dots are external to the system. (Friston's original image differentiated the sublayers of the blanket. For the record, since his colours are mentioned in the text: the external dots were pale blue, the internal ones were dark blue, the sensory ones were purple and the active ones were red.)

the core do not have any impact upon the outer sublayer). Likewise, the states of the inner sublayer are causally affected by those of the inner core, and they in turn influence those of the external milieu, but the line of influence is again not reciprocated. This arrangement of causal dependencies defines the properties of what is known as a *Markov blanket*.[37]

The 'Markov blanket' is a statistical concept which separates two sets of states from each other. Such formations induce a *partitioning* of states into internal and external ones, i.e. into a system and a not-system, in such a way that the internal states are insulated from the ones that are external to the system. In other words, the external states can only be 'sensed' *vicariously* by the internal ones as states of the blanket. Moreover, a Markov blanket is itself partitioned into subsets that are, and subsets that are not, causally dependent (directly) upon the states of the external set. These states of the blanket are called 'sensory' and 'active' states, respectively.

The formation of a Markov blanket thus divides a system's states into four types – internal, active, sensory and external – where the external states are not part of the self-organising entity. Crucially, the dependencies between these four types of state create a circular causality. The external states influence the internal ones via the sensory states of the blanket, while the internal states couple back to the external ones through its active states. In this way, the internal and external states cause each other in a circular fashion. Put differently, sensory states *feed back* the consequences of the effect on the external states of the active states, and thereby adjust the subsequent actions of the system.

If this sounds like the perception/action cycle in living organisms – and reminds you of what I said in the previous chapter about the circular causality between reticular activating system, forebrain and midbrain decision triangle – that is no accident. This is precisely the value of such abstract models. They reveal deeply regular formalisms that can then be recognised across a wide range of substrates – and enable us to understand the structure of living things in new ways.

Once you start looking, Markov blankets are everywhere. A cell membrane has the properties of a Markov blanket, and so does the skin and musculoskeletal system of the body as a whole – which is itself comprised of cells. So does every organelle, organ and physiological system. The brain (actually the entire nervous system) – which regulates the body's other systems – therefore possesses a Markov blanket. In fact, it is a *meta-blanket*, since it surrounds all the other blankets. Self-organising systems can always be composed of smaller self-organising systems – not all the way down, but certainly a dizzyingly long way. That is the basic fabric of life: billions of little homeostats wrapped in their Markov blankets.

We are closing in on an answer to the question I posed above: why and how, in physical terms, does question-asking arise? We are not quite there yet, but from what I have told you so far it is reasonable to conclude that the very selfhood of a complex dynamical system is constituted by its blanket. Such self-organising systems come into being by separating themselves from everything else. Thereafter, they can only register their own states: the not-system world can only be 'known' vicariously, via the sensory states of the system's blanket. I propose that these properties of self-organisation are in fact the essential preconditions for subjectivity.

Let me draw a distinction. It is the self-preservative nature of such systems, their tendency to separate from their environments and then actively maintain their own existence, that provides the elemental basis of *selfhood*. And it is the insulated nature of such systems, the fact that they can only register the not-self world via the sensory states of their own blankets, that constitutes the elemental basis of *subjectivity* – the 'point of view' of a sequestered self.

This does not imply that every self-organising system possesses *sentient* subjectivity, of course. We are still some way from being able to identify the specific properties that a self-organising system must display before it can become conscious. However, even without consciousness entering the picture, we do seem to have stumbled on a physical prototype for the problem of other minds. The very nature of a Markov blanket is to induce a partitioning of states into 'system' and 'not-system' ones, in such a way that not-system states are *hidden* from the system's interior, and vice-versa.[38]

Let us return to Friston's primordial soup, where things get stranger still. Having spontaneously generated a complex dynamical self-organising system, he tested whether this ensemble makes it possible to *predict* external states from the internal states of the system. If it does, Friston argued, this would suggest that the internal states of a system have *modelled* their external states over time, and could also be said to *represent* those external events within themselves. I know this sounds magical but it simply means the system has adjusted to patterns of external events; that it has accommodated itself to them. (To oversimplify the point: that is why you can predict the typical wind direction of an area from the slope of the trees within it when the wind is not blowing. The slope of the trees 'represents' the typical wind direction because they have grown at that angle to accommodate it.)

Friston examined the functional status of the internal subsystems of his simulated organism, and this predictive capacity is exactly what he found:

> The internal dynamics that predict [an external event] appear to emerge in their fluctuations before the event itself – as would be anticipated if internal events are modelling external events. Interestingly, the subsystem best predicted was the furthest away from the internal states. This example illustrates how internal states infer or register distant events in a way that is not dissimilar to the perception of auditory events through sound waves – or the way that fish sense movement in their environment. [The] subsystems whose motion could be predicted reliably [are] the most significant at the periphery of the ensemble, where the ensemble has the greatest latitude for movement. These movements are coupled to the internal states – via the Markov blanket – through generalized synchrony.[39]

Having observed this synchrony, whereby the internal states of the system modelled physically distant events, Friston concluded that the internal states exhibit 'inference'.

This turns out to be the most significant property of such systems. The Markov blanket endows the internal states of self-organising systems with a capacity to represent hidden external states probabilistically, so that the system can *infer the hidden causes of its own sensory states*, which is something akin to the function of perception. This capacity, in turn, enables it to *act purposively* upon the external milieu, on the basis of its internal states – which actions are akin to motor activity.

In this way, the system maintains and renews itself in the face of external perturbations.[40] Merely being a self-organising system is sufficient to confer a purpose on it and on each of its parts, and that is the function of the active states of the blanket: they manipulate the environment in order to maintain the integrity of the system. Which means that, along with an enclosed self, a subjective point of view, a goal and the capacity both to sense and act, the mere fact of a Markov blanket brings about something akin to *agency*.[41] It might not seem particularly commanding in the form in which it appears in Friston's simulations – the pale blue (external) dots influence the dark-blue (internal) ones via the purple (sensory) ones, while the dark-blue dots couple back to the light-blue ones through the red (active) ones. Nevertheless, I hope you can see why biological self-organising systems must infer the hidden causes of their sensory states – albeit non-consciously, for now. If they did not, they would cease to exist. They are obliged to model causal dependencies in the world, so that their actions in that world ensure their survival.

This is where the concept of 'expected states' comes from, and why biological self-organising systems are homeostatic. Homeostasis seems to have arisen with self-organisation. The sensory and active states of a Markov blanket are nothing other than a self-organising system's receptors and effectors, and the model of external states that it generates is its control centre.

Biological self-organising systems *must* test their models of the world, and if the world does not return the answers they expect they must urgently do something differently or they will die. Deviations from expected states are, therefore, a foundational form of Wheeler's 'equipment-evoked responses'. This is

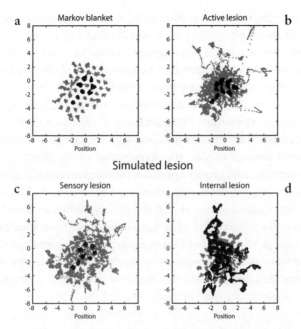

Figure 14 Entropic effects of slightly damaging the
Markov blanket of a self-organising system.

how question-asking arises; *self-organisation brings participant observers into being*. The question that a self-organising system is always asking itself is simply this: 'Will I survive if I do that?' The more uncertain the answer, the worse for the system.

The relationship between the active states of a Markov blanket and self-preservation, via homeostasis, is best illustrated by showing what happened when Friston *damaged* the blanket of the self-organising systems in his experimental simulations. He did this by selectively preventing the sensory states of the blanket from influencing its active states (see Figure 14, panels [b], [c] and [d]). In the absence of the usual active states of the Markov blanket, entropic chaos ensued and the system rapidly dissipated. That is to say, it ceased to exist.

What I have just illustrated in formal terms (using concepts from statistical physics) is the basic structure of any dynamical self-organising system. Friston summarises the lawful relations as follows: '*any ergodic random dynamical system that possesses a Markov blanket will appear to actively maintain its structural and dynamical integrity*'.[42] (An 'ergodic' system occupies limited states.) This type of activity meets Kant's criteria, stated above: a Markov blanket is both the end and the means by which a self-organising system persists over time – and this occurs naturally. Thus, such systems appear to have minds of their own, albeit very primitive (and non-conscious) ones.

If this argument is correct, then it should be possible to identify the emergence of self-organisation in any arbitrary ensemble of coupled subsystems with short-range interactions. That is precisely what Friston's primal-soup experiment showed. From his experiment, he therefore abstracted four fundamental properties of *all* biological self-organising systems:

(1) they are ergodic.
(2) they are equipped with a Markov blanket,
(3) they exhibit active inference,
(4) they are self-preservative.

What exactly is the Free Energy Principle which gives this chapter its title? Notoriously difficult to explain, for one thing. This is largely due to Friston's equations, which, as I discovered when I made myself understand 'Life as we know it', are both abstruse and opaque. I am therefore going to try to explain the principle to you entirely in words. Equations can always be translated into words because they are nothing other than statements of relationships. That said, since the basic equation explains your central aim and purpose in life, as well as that of everything else that has ever lived, you should probably see it in its canonical notation at least once. It goes like this:

$$A = U - TS$$

where A denotes free energy, U denotes total internal energy, T is temperature and S is entropy.

What does it mean? In thermodynamics, the free energy of a system is equal to the total amount of energy contained in the system minus the part of that energy that is already employed in effective work and therefore is not free.[43] The equation, therefore, just says: 'Free energy is equal to the total internal energy minus the energy already employed.' What could be simpler? Free energy is what is left when you take away the energy that isn't free (i.e. when you take away the 'bound energy').[44]

The equation I have just described accurately quantifies free energy in basic thermodynamic contexts, but not in chemical ones, where allowance needs to be made for the additional molecules that are formed by some processes at various temperatures and pressures. Chemists therefore use a slightly different version of the equation.[45] To differentiate them from each other, the classical thermodynamic type of free energy is called 'Helmholtz free energy' (for Hermann von Helmholtz, a leading member of the Berlin Physical Society) and the chemical-ensemble type is called 'Gibbs free energy'.

Friston uses a third version of the same equation, to quantify free energy in *information* contexts. I will call this type of free energy 'Friston free energy'. The relevant equation says that 'Friston free energy is equal to average energy minus entropy'.[46] Here, 'average energy' means the *expected* probability of an event happening under a model, and 'entropy' means the *actual* incidence of it happening. So Friston free energy is the difference between the amount of information you expect to obtain from a data sample – from a sequence of events – and the amount of information you actually obtain from it. (You will recall that the entropy of a predictive system is its average information, where increasing information means decreasing probability.)[47] The equation 'Friston free energy is equal to average energy minus entropy' therefore says basically the same thing as the equation 'Helmholtz free energy is equal to the total internal energy minus the energy that is not available to perform work'.[48] This is because Friston free energy is analogous to Helmholtz free energy, where there is an *information* exchange as opposed to a *thermodynamic* exchange between the system and its environment.[49]

If biological systems must minimise their entropy, and entropy is average information, then it follows that they must keep the flow of information they process to a minimum. They must minimise unexpected events. This is technically known as 'surprisal'. Like entropy, surprisal is a declining function of probability: as the probability goes down, the surprisal goes up.[50] Surprisal measures how unlikely an event was; entropy measures how unlikely it is expected to be (on average).[51] So, surprisal, like model entropy, is a bad thing for living organisms. I do not want us to get lost in the technicalities.[52] Let me just say that, at the most brutally basic level, you are in a surprising state if you move outside the set of states you biologically *expect* to be in (e.g. below or above 36.5–37.5 °C, or breathing underwater), precisely because there is a low probability of your being in that state.

Self-organising systems must minimise information flow, because increasing information demand implies increasing uncertainty in the predictive model. Uncertainty yields surprises, which are bad for us biological systems because they can be dangerous. But how can we minimise surprises by minimising information flow? Isn't that just putting our heads in the sand? The answer is no. Friston free energy is a quantifiable measure of the difference between *the way the world is modelled* by a system and *the way the world really behaves*. Therefore, we must minimise this difference. A system's model of the world must match the real world as closely as possible, which means that it must minimise the difference between the sensory data that it samples from the world and the sensory data that were predicted by its model. This will maximise 'mutual information' between the world and the model, which minimises uncertainty.

One way to do this is by *improving the system's model of the world*. Prediction errors can be fed back to the generative model, so that it generates better predictions next time. Fewer errors made means fewer errors to feed back into the model, which means less information flow. 'Mutual information' is therefore a product of communication, of question asking and answering: is the world behaving as I predicted, yes or no?

Now, since biological systems like you and I are insulated from the world by our Markov blankets, we cannot compare our models directly with the way the world really is. We must therefore bring the whole process of minimising surprisal inside our heads, and become both the 'source' and 'receiver' of the information that flows from our question-asking.

We do this by measuring relative entropies – by quantifying the gap between the sensory states predicted by an action and the sensory states that actually flow from that action. This yields the quantity called Friston free energy, which is always a positive value greater than the actual surprisal.

Sensory evidence (received in the form of spike trains generated inside our heads – in effect, billions of ones and zeros) are the only data we can get. From that data we must *infer* the causal structure of the world. Markov-blanketed beings that we are, we are obliged to rely on things like probability distributions rather than absolute truths. That is why it is useful to know that Friston free energy is always greater than surprisal; it enables our brains to approximate unknowable truths using statistical calculations.

As we saw in Friston's soup experiment, generative models come into being with self-organising systems. For that reason, they are sometimes called 'self-evidencing' systems, because they model the world in relation to their own viability and then seek evidence for their models. It is as if they say not 'I think, therefore I am' but 'I am, therefore my model is viable'.[53] The self-model of each biological system is determined in part by its *species*, as I have explained already. You – as a human being – will expect to find yourself in vastly different states from those a shark normally inhabits. It is highly improbable for you to find yourself respiring water hundreds of metres below the surface of the ocean, but not at all improbable for a shark. So a given set of sensory states is more or less surprising – more or less improbable – depending on which species of organism is in them.

The test of a good model of the self-in-the-world is how well it enables the self-system to engage the world in ways that keep it within its viable bounds. The better these engagements are, the lower its free energy will be. The lower its free energy, the more of the system's energy is being put to effective, self-preserving work.[54] The Free Energy Principle thus explains in mathematical terms how living systems like you and I resist the Second Law of Thermodynamics through homeostasis-maintaining work.

It also explains another way in which self-organising systems are self-evidencing (one wants almost to say introspective): they are obliged to ask questions *of themselves* about their own states. Specifically, they are chronically obliged to ask: 'What will happen to my free energy if I do that?' The answer to this question will always determine what the system does next, over a suitable time period.[55] This is the causal mechanism behind all voluntary behaviour.[56]

Surprisingly, we have not yet got to the end of the remarkable mental powers on display in Friston's primordial soup. We have seen how even the most basic self-organising systems have non-conscious selves and subjectivity, agency and a purpose (namely, to survive). They perceive and, as necessary, they act. To this impressive list we can add yet another prize. It turns out that they exhibit a kind of rationality. They are, to a decent approximation, Bayesians.

The Reverend Thomas Bayes was an English clergyman and theologian whose name survives entirely thanks to his insights into probability, which he never bothered to publish. It was Bayes who taught us (in a paper published posthumously in 1763) that we should use *current evidence* in conjunction with *background knowledge* to make and revise our *best guesses* about the world. In other words, using Friston's now-familiar terminology, we should take sensory samples, compare them with predictions derived from our generative models, and update our beliefs accordingly.

Here's the standard expression:[57]

$$P(A \mid B) = \frac{P(B \mid A)\,P(A)}{P(B)}$$

Translated into words, this theorem says: 'the ratio of probabilities for two hypotheses conditional on a body of data is equal to the ratio of their conditional probabilities multiplied by the degree to which the first hypothesis surpasses the second as a predictor of the data'.

More simply: given a hypothesis followed by some evidence, you must revise the probability of the hypothesis by considering its *likelihood* in conjunction with its *prior probability*. The 'likelihood' of the hypothesis is the degree of fit between what it predicts and the evidence actually obtained, and the 'prior probability' is your background knowledge about the hypothesis (i.e. its probability even before you considered the new evidence). The result is the *posterior probability* of the hypothesis. Given two competing beliefs, then, your best guess is the one with the highest posterior probability. (See Figure 16 on p. 199.)

For example: you are at Cape Town airport and you see someone alighting from a Johannesburg flight who looks like your friend Teresa. Your hypothesis is that this is Teresa. Its 'likelihood' is the probability that you are seeing someone who looks like this, given that she is Teresa. Then you remember that Teresa lives in London, which reduces the 'prior probability' that this is her. Your conclusion (the 'posterior probability') is that you are looking at someone else who merely resembles Teresa.[58]

The most important thing about Bayes's theorem for the purposes of neuroscience is that it explains how perceptual inference – an unconscious process – actually works in real life, and how signal transmission actually works in real sensory-motor processing. Brain circuits literally *do* compute prior probability distributions and then send predictive messages to sensory neurons, in an endless effort to dampen the incoming signals; and perception literally *does* involve comparisons between the predicted and actual distributions, resulting in computations of posterior probability. The resultant inferences are what perception *actually is*.[59] Perception is an endeavour to self-generate the incoming sensory signals and thereby explain them away. That is why so many neuroscientists nowadays speak of the 'Bayesian brain'.

Remember what I said earlier about the relationship between 'information' and the material world: even though you cannot see and touch information, as such, there is no question as to whether it really exists. The behaviour of physical systems is *determined* by information flows. Hence, minimising Friston free energy simultaneously minimises Gibbs and Helmholtz free energy. This is because minimising prediction error minimises information flow, and reducing information flow reduces *metabolic expenditure* on the part of the brain and the body as a whole.[60] This is not only because brain activity burns up so much energy (20 per cent of our total supply). It is also because the minimisation of statistical free energy in the brain regulates physiological energy exchanges between the body and the world.[61] These exchanges, too, entail metabolic expenditure. The predictive brain is thus revealed to be 'lazy' (over the long term): vigilant for every opportunity to achieve more by doing less.[62]

This is a minimalist explanation of what brains do. However, in order to keep the organism alive, brains must do more than conserve energy resources; they must also take account of the host of other biological needs (in addition to energy balance) that I discussed in Chapter 5, almost all of which oblige us to perform work in the outside world. The multiplicity of needs that characterise us complex organisms, you will see, has everything to do with consciousness.

We have learnt that the suppression of prediction error is the essential mechanism of homeostasis. Therefore, minimising free energy becomes the basic task of all homeostatic systems. Friston's free-energy equation turns out to be a reformulation, in quantifiable terms, of Freud's definition of 'drive': 'a measure of the demand made upon the mind for work in consequence of its connection with the body' – a measure that Freud considered to be unquantifiable. Now we can quantify it. The fundamental driving force behind the volitional behaviour of all life forms is that they are obliged to minimise their own free energy. This principle governs everything they do.

As Friston pithily formulated it, the Free Energy Principle dictates that all the quantities that *can* change – i.e. that are part of the system – *will* change to minimise free energy.[63] In fact, this is more than a pithy formulation; it is a law. Let's call it Friston's Law: All the quantities in a self-organising system that can change will change to minimise free energy. So, there you have it. Armed with this knowledge, everything that we call mental life becomes mathematically tractable.

We are almost ready to learn what consciousness is, in formal and mechanistic terms. But first, a fable.[64]

8

A Predictive Hierarchy

A structural engineer, Eve Periaqueduct, is employed to prevent and repair leaks in a municipal dam. She doesn't know that the deeper purpose of her job is to ensure reliable water and power for a nearby village, and to keep it from being swept away by floods. But she doesn't need to know; her sole task is to minimise leakage in the dam. (She might also remember from her university studies that she is minimising the entropy of the dam. But she doesn't need to remember that; her job is very practical.)

Her employers supply her with the equipment she needs, together with a small team of workers. She also inherits a set of instructions drawn up by her predecessors outlining where the weakest points in the dam are – telling her what to do and when. She and her team diligently maintain and repair the dam, proactively focusing on its weak points while they also plug any spontaneous leaks as they appear. Gradually, over years, she learns that some of the unexpected leaks, too, follow regular patterns. She therefore *updates* the instructions handed down to her, becoming more adept at predicting (and therefore preventing) leaks. This saves on costs.

It dawns on the enterprising Eve that the long-term leakage patterns she has recorded correlate with climatic conditions. Her records have unwittingly *modelled* the local climate (i.e. her records and the weather carry 'mutual information'). Without trying to, she has generated a model of an aspect of the world beyond the dam. There are patterns in the weather, and they correspond to patterns in the leaks.

Building on this insight, Eve hires additional staff to establish a meteorology department, which she calls her 'weather-sensing' department. This yields a new tier in the hierarchy of her team, located at a different site, justified by the expectation that having better weather forecasts will save on repair costs in the long run.

The new tier makes her predictive model more sensitive to expected contexts. The weather-sensing staff do not need to know that their job has anything to do with preventing leaks; they focus solely upon the task of predicting changes in the weather. Eve supplies them with a chart of expected conditions derived from the inherited instructions she began with, as updated by her. Note that these instructions are not about expected leakage patterns but rather about expected climatic conditions.

Because Eve doesn't want to waste a lot of time checking messages, the new department is asked to send her feedback reports only if *deviations* from these expected conditions occur. She calls these 'error' reports and uses them to further update her chart of expected climatic conditions, which she in turn sends back to the weather station, in the knowledge that this will reduce their work load – and ultimately her own.

All of this enables the meteorology department to focus efficiently upon the task she has given them. To fulfil their duties, the department installs a series of weather-sampling instruments at various locations, some located at great distances from the dam. These barometers, thermometers, precipitation meters and the like are calibrated by the team in such a way that they only send signals to the weather station when the parameters they sample (air pressure, temperature, humidity, etc.) deviate from expected bounds. These bounds are set in accordance with the predicted climatic conditions. This, again, saves costs, as the workers who are hired to read the meters (creating a further tier in the hierarchy) only need to visit those instruments that transmit 'error' signals to the weather station. By keeping careful records of these signals, the station is enabled to adjust periodically the expected bounds for each instrument, thereby further *automatising* their procedures (i.e. reducing the frequency of the signals which require them to send out meter readers and adjusters).

Some of the resultant algorithms become rather subtle, as the team learns that fluctuations in the parameters they measure are not necessarily fixed and regular; they vary on a *contextual* basis. For example: 'If air pressure is down, increase expected precipitation, but only in winter.' Even so, the meter readers and adjusters do not need to know anything about the greater task of the meteorology department. Their sole job is to read meters and adjust instruments, in accordance with the updated instructions they receive from the meteorology office. Even less relevant to their job is the fact that the meteorology department's reports concerning deviations from expected climatic conditions are sent to Eve, to enable her to more reliably predict patterns of leaks, and thereby undertake more efficient dam maintenance work.

Incidentally, the nearby villagers make use of these weather forecasts for their own purposes, which have nothing to do with the dam. This gives them a wrong-headed impression of the real purpose of the department; they believe it is there to help them schedule outdoor social activities.

Eve Periaqueduct notices over time that the pattern of dam leakages correlates not only with climatic conditions but also with seismic events. She therefore establishes a second dedicated team, which she calls the 'earthquake-sensing' department. This seismology department focuses solely on modelling and predicting tectonic shifts and the like, which results in the second department installing, calibrating, monitoring and continuously adjusting technical 'sensory' equipment of its own. It also compiles complex records – which, as happened with the meteorology department, and with Eve herself, enable the new team to automate aspects of their work and to focus only upon unpredictable, short-term fluctuations. (The villagers make good use of these forecasts, too, of course, although that was never the point of them.)

What thus gradually emerges is a complex predictive hierarchy, with multiple departments, each with sublayers of their own which sample different parameters in the world beyond the dam. Each level in the hierarchy follows only the updated predictive instructions it receives from the level above, and reports only deviations from expected states in the parameters below which are monitored at that level. From Eve's point of view, the composite reports she receives from her sensory departments contextualise each other. She has to decide, from time to time, which report to *prioritise*. After all, she has limited resources; she cannot cover all possible events.

Eve still maintains the dam on the basis of her long-term schedule, from which she deviates only when it does not fit with the combined forecasts she receives from her sensory departments. These departments, in turn, only send feedback reports to Eve when the data samples they gather deviate from their own long-established predictions. And so it goes on, down to the meter readers and adjusters.

Incidentally, all this iterative message-passing and schedule-updating between the tiers of the organisation that Eve established follows Bayes's rule: it uses current evidence (sensory samples) in conjunction with background knowledge (prior hypotheses) to make *and revise* its best guesses (posterior hypotheses) about the world.

As time passes, Eve's work becomes repetitive and boring, and she looks forward to retirement. She finds herself thinking, 'Before I go, I would like to build an entirely new and better dam'. So, she calls up the municipality and asks: do we happen to have a 'reproductive' department?

A lot of quasi-mental functions emerge with the most basic self-organising systems, but to explain how real brains work, we need to look at how these systems can fit together into an overarching, mutually beneficial structure. It has become evident, thanks to Friston's work, that the nervous system implements an iterative predictive hierarchy that works very much like the one that Eve Periaqueduct established over time. The brain's many complex functions really can, ultimately, be reduced to a few simple mechanisms like this. As Jakob Hohwy writes:

> The brain is somewhat desperately, but expertly, trying to contain the long and short-term effects of environmental causes on the organism in order to preserve its integrity. In doing so, a rich, layered representation of the world implicitly emerges. This is a beautiful and humbling picture of the mind and our place in nature.[1]

Please note, that in cognitive science the word 'implicit' means *unconscious*.

At the core of the brain's model of its self in the world, species-specific predictions about its viable bounds are generated (the left side of Figure 15).[2] These are embodied in autonomic reflexes which take the form: 'if I do that, my temperature will be approximately 37 °C.' At the next level (to the right), surrounding this core, the brain generates instinctual behaviours (which take the form of the innate predictions that I described in my review of the basic emotions). At the next level, it generates acquired involuntary behaviours (from its non-declarative long-term memory systems). At the next, it generates voluntary behaviours (from its declarative long-term memory systems). And at the final, outermost level, it generates the most tentative, here-and-now actions that 'predict the present' (from its short-term memory systems).

I am simplifying, of course: there are many more than five levels in the brain's predictive hierarchy and they are arranged in multiple, parallel processing streams. All the same, some general principles emerge.

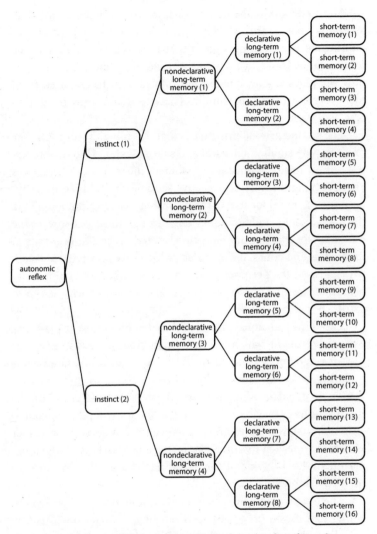

Figure 15 A simplified predictive hierarchy, extending from the tonic autonomic core to the phasic sensory-motor periphery. Predictions flow from the core to the periphery (from left to right in this diagram); prediction errors flow in the opposite direction.

The first is that the brain conspires to anticipate and thus 'explain away' events in the world. It suppresses predictable, uninformative incoming signals that it would otherwise have to process pointlessly. In short, each level in its hierarchy receives only the newsworthy, unexpected information transmitted from the level immediately beyond it. These feedback reports are prediction errors.

The second general principle is that this hierarchy unfolds over progressively smaller temporal and spatial scales. The core predictions apply in all circumstances, whereas the more peripheral ones are fleeting and focal. A predictive sequence unfolds from body-monitoring nuclei located in the brainstem and diencephalon,[3] via the basal ganglia and limbic system, through the neocortex, to the modality-specific sensory receptors located in the end-organs (e.g. in the rods and cones of the retina), which have very narrow receptive fields. At the periphery, short-term accuracy and complexity prevail at the cost of long-term generalisability, which is enjoyed by the deeper predictions.

The third principle is closely related: a hierarchy of plasticity exists in terms of which the core predictions cannot change but the peripheral ones can and do; they are subject to instantaneous updating, with the intermediate levels yielding an intermediate degree of plasticity. This means that the 'control centre' of the brain's homeostat (its self-model), constantly updates itself – although it shows increasing resistance to change as the cascades of errors approach its core. The increasing plasticity of the more peripheral levels is one of the major advantages of a *hierarchical* predictive model.

The fourth principle, which I have not made clear before now, is that perception (as opposed to learning) *reverses the direction* of information processing. By inverting the causal dependencies that shaped the predictive model in the first place, the brain produces our perceptual inferences – which Merker described as a 'fully articulated, panoramic, three-dimensional world composed of shaped solid objects: the world of our familiar phenomenal experience'.[4] (These inferences flow from left to right in Figure 15.) 'Inversion of the predictive model' simply means shifting from learning to predicting on the basis of what you have learnt. This is what Eve Periaqueduct did; she inferred ('perceived') the state of the world beyond her dam from the meteorological and seismological data she received.

Perception proceeds *from the inside outwards*, always from the viewpoint of the subject. It really is apperception, an inferential process, a matter of Bayesian hypothesis-testing.[5] Hermann von Helmholtz, who was the first to grasp the essentials of this, called it 'unconscious inference' (again, note the adjective). What you see is your 'best guess' as to what is actually out there; it is *your proposed answer to the questions you are currently putting to the world*.

The brain must infer the most likely causes of its incoming signals without any direct access to the unknowable world beyond its blanket. All that the brain has to go on is the way that its own sensory states (the spike trains; Figure 11) flow and alter. Its task is to use these signals to create a probabilistic *model* of the regularities that exist in the real world (or rather, between itself and the world) which it then uses to generate inferences that guide its actions – actions which must ensure its survival in that world. The actions, in turn, generate new sensory samples, which are used to further update the model, which it must do because models are imperfect things. This leads to new actions; and so on.

Actions should therefore be viewed as *experiments* that test hypotheses arising from the generative model. If an experiment does not yield the predicted sensory data, then the system either (1) must change its prediction to better explain the data, or, if it remains confident about the original prediction, (2) must obtain better data; that is, it must perform actions that will change its sensory input.

These two options – changing the prediction or the input – are the fundamental mechanisms of perception and action respectively.

The three paragraphs above prompt me to correct a bias that has characterised this chapter and the preceding one. Until now, like most scientists who use cortical functions as their model example of how the brain works, I have focused almost exclusively on Bayesian *perceptual* inference. But there is also *active* inference. In fact, active inference is the primary form (at least in biology), since the whole point of perception is to guide action.

As I have just said, the Bayesian brain has two ways of responding to prediction error. When faced with a hypothesis to which decreasing 'posterior' probability applies, it creates a better fit between the hypothesis and the data by changing either its 'prior' prediction or its input. The difference between these two alternatives comes down to the statistical *direction of fit*: error is reduced if the prediction is changed to match the sensory input, and it is also reduced if the sensory input is changed to match the prediction. In reality, of course, organisms alternate between these two options all the time. (Think of a field mouse darting through the scrub, stopping to look around, darting again, stopping to look around again, and so on.) In some respects, perception and action are more similar than they seem.

The body itself is a hidden world that is 'external' to the Markov blanket of your central nervous system. The cascade of predictions in Figure 15 could have included equivalent concentric layers culminating in the terminal visceral receptors and effectors which operate your *internal* organs. So the brain's model of the world must include a model of your bodily self and its trajectory no less than it does of all the other hidden causes of interest to you. (In my earlier example, you could not have escaped the smoke-filled room if you did not have an implicit model of how your body moves *and* how it balances blood gases.)

Action, furthermore, does not come about by the brain broadcasting some grand plan to all the muscles and organs of the body. What happens instead is that the muscles contract and the glands secrete until the prediction errors they transmit through the hierarchy disappear. Thus, the musculoskeletal and visceral 'action' organs of the body are at the mercy of the error signals that are generated by differences between what the predictive model expects them to achieve and what they actually achieve. Suppressing prediction error is what controls action, no less than perception.[6]

Remember Friston's Law: all the quantities in a self-organising system that can change will change to minimise free energy. The multiple bodily homeostats regulated and orchestrated by the midbrain's meta-homeostat are the pivot of the mechanism by which we stay alive, for the simple reason that homeostatic regulation maintains our bodies within their viable bounds. These bounds cannot be changed. This means – in accordance with Friston's Law – that *something else in the system must change*. This is the formal, mechanistic explanation of the imperative link that exists between drive and *action*, and it is why there must be *a hierarchy of prior predictions*, some of which can be changed and some of which cannot.[7]

Yet blind action is of little use. It must be guided by perception, which is generated by a model of the self in the world. Bayes's rule describes how the predictive model that achieves this is implemented, and how it is thereby constantly updated, and why it must be updated. This becomes the formal basis of *learning*, whereby predictions are acquired and nuanced over time from incoming error signals. These system dynamics also confirm, on equally mechanistic grounds, that action takes priority over perception: it is only action that can increase the probabilities of prior predictions – some of which, as I said, simply cannot be changed.

Bayes's rule starts from an assumption of what we called 'background knowledge'. The rule cannot work otherwise.[8] This raises a question: where does the background knowledge come from, at the outset, before the system has gathered any evidence about the world? The answer is that our core 'expected states' are encoded by our species as *innate* homeostatic settling points – quantities that were determined by what worked effectively for our evolutionary ancestors. We are beneficiaries of the biological successes of past generations, which fix the most basic premises of our existence.

And we cannot rest on their laurels. The connection between affect and action dictates that if at first you don't succeed, you must try, try and try again. The demands of your deepest biological drives are inexorable: they can be quieted by only satisfaction or death. Short of the latter, you must supplement the reflexes and instincts you were born with and develop *other* ways of meeting your needs. There is no alternative. In other words, you must learn from experience. Fortunately for us, the human brain is unusually well equipped to do that.

An interesting implication of all this is that, if affect does indeed work by the homeostatic mechanisms I have described – if it really is 'a measure of the demand made upon the mind for work in consequence of its connection with the body' – then it must be *the* fundamental vehicle of free-energy minimisation. Affect, therefore, is the primary medium of volition – and the fount of all mental life.

I said above that learning from experience produces a hierarchical model of the world that, when inverted, generates predictions about that same world. But the process does not end there; predictions must be *tested*. This causes prediction errors, which are used to update the model. That is what learning 'from experience' is all about. Prediction errors are the sensory signals that were *not* predicted by a current hypothesis, i.e. the ones that were not self-generated. This is the salient part of the data.

The mistake that most cognitive scientists make at this point is to assume that the incoming data is exclusively *exteroceptive*. They forget that the prediction errors (sensory inputs) that matter most to us come from *within*. Deviation from expected core body temperature, for example, provides 'sensory' feedback, no less than unexpected external events; and so does the homeostatic error signal that gives rise to suffocation alarm. These signals generate *affects*, not perceptions. As Freud said, the forebrain is a 'sympathetic ganglion'. Confusion on this score is the perennial price my colleagues pay for adopting the cortical fallacy.[9] Consciousness is endogenously generated; all of it. Consciousness at its source is affect. Then it is extended outwards onto perception, to evaluate perceptual inferences, in the manner I shall now describe.

Finally, we can address the question: why and how do the natural self-preservative functions described in this chapter *become conscious*? We know that consciousness is grounded in affect, in feelings. But what are the formal, mechanistic laws that give rise to feelings – and thereby to consciousness?

9

Why and How Consciousness Arises

The basic question that living things must always ask themselves is: 'What will happen to my free energy if I do that?' Do what, though? The potential courses of action at any given moment are not arbitrary or infinite; they are dictated by current *needs*. There is, of course, an intimate link between needs and actions; each need demands its own adequate action. If you are hungry you must eat. If you are tired you must rest. However, there is an executive bottleneck: you can only do one or two things at once. This means that, to select your next action, you must rank your current needs by urgency.

Two things make this ranking task more complicated than it might initially appear. First, the needs of us complex organisms don't usually have to be satisfied in a fixed order. Eating and sleeping – which is more important? It depends on all sorts of considerations. And, second, many of the needs of complex organisms cannot always be satisfied by the same action. Eating soup requires different skills from eating corn on the cob, for example (and don't even think about the skills and resources involved in preparing them). In both of these respects, 'what to do next' chronically depends upon context. That is why internal needs must be prioritised in relation to *prevailing external conditions*.

We are born with species-specific predictions about what to do in states like hunger, thirst, fear and rage. These innate predictions are called 'reflexes' and 'instincts' – inherited survival tools for which we have every reason to be grateful. But they are not flex-ible enough to deal with the range and complexity of situations that we actually contend with; so they must be supplemented. This is the role of learning from experience.

We have already linked learning from experience with the Law of Affect. Affective valence – our feelings about what is biologic-ally 'good' and 'bad' for us – guides us in *unpredicted* situations. We concluded that this way of feeling our way through life's unpredicted problems, using *voluntary* behaviour, is the biological function of consciousness. It guides our choices when we find our-selves in the dark. But of course, for it to be able to do that, it must link our internal affects (rooted in our needs) with representations of the external world.

This explains why arousal is accompanied by feelings *and* con-scious perceptions of things. I conceded at the end of Chapter 6 that the great mystery of this conjunction – the mystery of how subjective experience fits into the fabric of the physical universe – could be solved only if we reduce physiological and psychological phenomena alike to their underlying mechanistic causes. These causes were to be revealed at a depth of abstraction that only *physics* could provide. In the previous two chapters I embarked on a formal account of those unifying mechanisms.

Now it is time to complete that account. If self-organisation and homeostasis do not by themselves explain why and how consciousness arises, what does? How, formally and mechanis-tically, does the biological *need-prioritisation* process I have just summarised relate to free-energy minimisation? And how does it happen that the outcome of this process *feels like something* to and for some self-organising systems?

The starting point of my answer is precisely the fact I just empha-sised: complex creatures like us vertebrates have multiple needs. That is, we have multiple internal subsystems, each regulated by their own homeostatic mechanisms, all of which contribute error values to the overall calculation of free energy. Our biological needs are these error values. When the needs become felt as affects, we describe them as positively or negatively 'valenced'. This means they possess subjective *value*: they feel good or bad to us. The behaviourists tried to objectify value by redefining pleasurable and unpleasurable feelings as rewarding and punishing stimuli; but we have already dealt with that issue. The valence does not reside in the stimulus; it is inherently subjective and qualitative. What is thrilling to one person is terrifying to the next.

But is it *impossible* to quantify valence? Consider Figure 12; the further the arrow deviates to the right, the greater the unpleasure. Thus, at any given moment, your hunger value might be 3/10 (which is worse than 1/10) and your thirst value might be 2/10 (which is better than 5/10), etc. Affective scientists take these sorts of measurements all the time; they ask research participants to rate their pleasures and unpleasures on what are called Likert scales. Although these scales are subjective, the fact that they are quantifiable *in principle* leaves the question open: why must affects be qualified? If self-organising systems can register 'equipment-evoked responses' (i.e. their own states) as quantities, in principle, then what does the quality add? This question concerns what philosophers call 'qualia' – the elusive mind stuff that supposedly cannot be accommodated within our physicalist conception of the universe.

The answer starts from the fact that needs cannot be combined and summated in any simple way. *Our multiple needs cannot be reduced to a single common denominator*; they must be evaluated on separate, approximately equal scales, so that each of them can be given its due. You cannot simply say that '3/10 of hunger plus 1/10 of thirst equals 4/20 of total need', and then try to minimise the total sum, because each need must be satisfied in its own right. Energy metabolism is not the same as hydration is not the same as thermoregulation, and so on; each of them is essential. As the behavioural neuroscientist Edmund Rolls puts it: 'If food reward were to *always* be much stronger than other rewards, then the animal's genes would not survive, for it would never drink water.'[1]

Taking these factors together, it makes sense for biological self-organising systems to distinguish their needs (their error values) *categorically*. The distinction between categorical variables is *qualitative*. Since error type *A* of 8/10 cannot be equated with error type *B* of the same value, for the reasons I have just explained, they must be treated as categorical variables. This enables the system to give each of them its due in the long run *and* to prioritise them contextually. That is why it makes sense for complex self-evidencing systems to categorise (to 'colour-code' or 'flavour') their multiple homeostats, so that they may compute them independently of each other and prioritise the outcomes.

Not only do the different needs contribute different quantities to total free energy; the different quantities also have different implications for the animal in different contexts (for example, hunger trumps sleepiness in some situations but not others). This contributes greatly to uncertainty – the mortal enemy of prediction machines. Increasing uncertainty is a dangerous state of affairs for any self-organising system: it predicts the system's demise. More uncertainty demands more computational complexity (which means more information flow, which means more entropy). Thus, categorisation becomes a necessity when the relative value of different quantities *changes* over time (if 8/10 for *A* is currently but not always worth more than 8/10 for *B*).

It is conceivable that an extremely complex set of model algorithms could evolve to compute relative survival demands in all predictable situations, to enable us *automatically* to prioritise actions on this basis. However, such complex models are extremely expensive, in every sense of the word. They are unwieldly, which means delay, which can be the difference between life and death; and they require lots of processing power, which means having to find more energy resources. Statisticians call the exponential increase in computational resources necessitated by a linear increase in model complexity the 'combinatorial explosion'.

Moreover, a complex model which accurately predicts what happens in one specific situation is unlikely to predict with equal accuracy what will happen in other situations. In statistical terms, we say that excessively complex models 'overfit' a data sample. Eve Periaqueduct's model of predicted leakage in her dam was not based on hour-by-hour and day-by-day events over the previous few weeks. Rather, it was based on long-term averages, drawn from many data samples collected over several years. This made her model simpler, and therefore more *generalisable*. On the principle of Ockham's razor (the Law of Parsimony),[2] we want *simple* predictive models. Simplification is essential if our models are going to apply in a wide range of situations. They must be serviceable, not only here and now but also in many other contexts.

So, predictive models must be simple. However, as Einstein famously said, 'Everything should be made as simple as possible, but no simpler.'[3] How does one strike the right balance? Compartmentalisation is the standard statistical method used to achieve optimal balance between complexity and accuracy. This takes many guises. For example, one part of your visual brain computes *what* you are looking at while another part computes *where* it is; this enables you to assume the constant identity of something while it moves around you, changing in shape, size and orientation. But what matters most is that a capacity to compartmentalise enables the system to rank its needs and their attendant predictions (i.e. the salient sources of expected free energy) categorically, over time, and to focus its computational efforts upon the prioritised compartment.

This is the statistical-mechanical basis for the observed fact that each affect possesses not only a *continuous* hedonic valence (a *degree* of pleasure and unpleasure, which is something common to all affects) but also a *categorical quality* (so that, for example, thirst feels different from separation distress, which feels different from disgust, and so on). These are the essential features of affective qualia, the elemental form of all qualia: they possess both quantity *and quality*. To state it more fully: affects are always subjective, valenced and qualitative. They have to be, given the control problem they evolved to handle.

It is useful to describe affective category selection and prioritisation in terms of operating modes. Think of the way an aeroplane behaves when it is in 'take-off' vs 'cruising' vs 'turbulence' vs 'landing' modes. The same variables come into play in these different situations, but they must be *weighted* differently each time. For example, your exact altitude matters a great deal more during landing than while cruising. The same applies to Eve Periaqueduct's dam. Different operating schedules were in force during her winter vs summer modes and during earthquake vs non-earthquake modes. Under earthquake conditions, Eve would be obliged to override the normal, automatised seasonal schedule and implement an 'earthquake emergency' schedule.

In the physiological terms I used in Chapter 6, the different operating modes are *state* functions of the brain. As I explained there, the midbrain decision triangle selects affective brain states – like 'suffocation alarm' mode. It does so when the PAG, the periaqueductal grey, answers the question: 'which of these converging error signals (i.e. needs) provides the greatest opportunity for minimising my free energy?' In other words, which need is the most salient *right now*? The answer is provided not only by the relative magnitudes of the competing error signals, but also by the differences between the categories (modes or states), the salience of which must be assessed *in context*. The contextual information, I explained, is provided by the superior colliculi.

Here is an example I noticed today. When I went for my jog at 7 a.m. it was dark, and when I returned an hour later it was light. (It is winter and I am staying in rural Sussex, writing this book.) Leaving, I passed a field adjacent to the farmhouse where a flock of sheep noticed me and they almost fell over each other to get away. Passing the same field on my return, the same sheep, lying in the same place, barely looked at me. Their startlement in the context of darkness was replaced by boredom in daylight. In short, the context altered the significance of the event 'human running towards me'. At night, this event is prioritised, which snaps the sheep into FEAR mode; by day it is not, and they remain in default-mode SEEKING.

This sort of thing determines what a system does next. In other words, it determines which *active* states will be selected by the generative model to resolve the prioritised category of uncertainty. It is as if the system says: under present conditions, *this is the category of prediction-error processing in which computational complexity cannot be sacrificed*. So, the system (in this case, the sheep) shifts into the FEAR operating mode. It executes the best strategy its generative model can supply in the circumstances: it flees. Then, after taking into account everything it has learnt about the particular field it is in (the expected context), it hopes for the best but prepares for the worst. By this I mean it cautiously executes its plan, ready to adapt to unfolding developments.

Crucially, shifting into FEAR mode means that the prioritised need has become an affect. In other words, it has become conscious. Why? It becomes conscious so that *deviations from expected outcomes* in the most salient category of need will be felt throughout the predictive hierarchy. That is what affect is. It is the 'equipment-evoked response' to the question the system asked of itself: 'which of these converging error signals provides the greatest opportunity for minimising my free energy?'

In South Africa, where I live most of the time, one has plenty of opportunities to witness how affect-selection works under natural conditions; that is, under the sorts of conditions in which this mechanism evolved in the first place. I am not talking about what goes on between lions and springbok in our fabulous nature reserves (although they, too, provide ample opportunities). I am referring to what goes on in our highly unequal and therefore troubled society. Many of my compatriots know what it feels like when the most salient need is to escape someone trying to kill you. At that point, your voluntary behaviour becomes dominated by feelings of FEAR, which gauge the here-and-now success or failure of your unfolding actions. Other needs (like the need to urinate) are then relegated to automaticity. In other words, you might wet yourself and you won't give it a second thought.

We are trying to reduce the psychological and physiological phenomena of affective arousal to a set of mechanical principles that can be formalised mathematically. In Chapter 6 I explained that once the midbrain decision triangle has prioritised a need, the forebrain model of its self in the world generates an expected context in which that need will be met. I said that there are two facets to this expected world: on the one hand it represents the actual content of our predictions, and on the other hand it must encode our level of confidence in those predictions. The first of these facets is supplied by the forebrain's long-term memory networks, which filter the present through the lens of the past. I introduced the principles governing this in Chapter 7. The second facet – the adjustment of confidence levels – is modulated by 'arousal'. So, let's formalise the laws that govern that.

The first mechanism I have identified so far is the one by which the most salient category of action (the most effective operating mode or state) is selected. That is how particular affective qualities first come to regulate the actions of complex self-organising systems. This results in the generating of sensory-motor plans, after which the system 'hopes for the best but prepares for the worst'. And this is the part of the story I want to focus on now. What is the causal mechanism by means of which hoping for the best and preparing for the worst is regulated?

The first part of the answer is that confidence levels attaching to predictions are *learnt through experience*, just like everything else. Then our level of confidence in our predictions can be predicted, just like the predictions themselves were. Predictive coding requires us to assign probabilities to the sensory states that we expect will follow from particular actions, and then compare those probabilities to the distributions actually observed in the ensuing sensory samples. This is the essence of the Bayesian 'hypothesis-updating' method I described earlier – the method by which we minimise our free energy.

Now, to determine the fit between a model and some data, it is not enough to simply compare the *means* of the distributions; it is necessary also to assess the *variation about the means* (see Figure 16). A large amount of variation in a sample makes one less confident about the fit. If the news reports: 'the King is dead … the King is dead … the King is dead', I am more likely to take it seriously than if it reports: 'the King is dead … no, the King is not dead … actually, it seems the King might be dead after all'. Judgements of difference between a predicted distribution and a data sample are easier to make when the distribution is narrow and *precise*.

What we must aim for is *precision* in our interactions with the world. Therefore, our models must possess a mechanism for *predicting precision*. This enables us to 'weight' the expected precision of incoming error signals relative to the precision we assigned to an outgoing prediction. This (relative degrees of confidence) will dictate the influence of the actual error signals over our predictions. If we are increasingly confident about an incoming error signal it should make us decreasingly confident about our current action plan, whereas vague unexpected developments should not deter us from our predetermined course. (Relative confidence values can be assigned to active versus perceptual expectations and to exteroceptive versus interoceptive ones, too, and to all the other quantities implicated in Friston's Law.)[4] This is a second-order kind of Bayesian inference, which entails *inferences about inferences*, i.e. educated confidence levels about predictions.

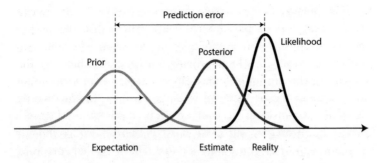

Figure 16 A prior prediction (on the left) is compared with a sensory data sample (on the right), resulting in a posterior prediction (in the middle). The 'means' of these three distributions are indicated by the vertical (dotted) lines, and their 'variance' is indicated by the horizontal (arrowed) lines. The wide expected variance in the prior distribution (horizontal line on the left) indicates a low degree of confidence in the prior prediction, and the narrow actual variance in the sensory data sample (horizontal line on the right) indicates a high degree of confidence in the data. In this example, the precision (inverse variance) in the sensory data is high; as a result, the posterior prediction shifts decisively to the right. If the precision in the sensory data were lower, the posterior prediction would shift less, if at all.

The purpose of precision modulation is to ensure that the inferences made by predictive models are driven by *reliable* learning signals (trustworthy news): if there is high confidence in a signal then it should be allowed to revise a prior hypothesis, and if there is low confidence then it should not. Confidence affects the strength of the error signals that are propagated inward through the hierarchy. A signal in which you have high confidence (i.e. a more precise signal) is a 'louder' one. It will therefore have a greater chance of delivering some residual error to the core of the system and a greater chance of updating its generative model. Conversely, less precise signals – signals in which you have less confidence, also known as 'noise' – can be sequestered in the sensory epithelium and (one hopes) safely ignored.

This means that we must minimise precise error signals. Again, that sounds paradoxical, until you realise that it just means we must avoid making glaring mistakes. The only way this can be achieved is by improving our generative models, thus increasing the mutual information between our models of the world and the sensory samples we obtain from it. In other words, we must maximise the precision of our predictions and then seek precise confirmatory data. *We must maximise our confidence in the beliefs that guide our actions.* This is called 'precision optimisation'.

We achieve precision optimisation by learning from experience. We must learn which sources of news we can trust (and when) and adjust our predictions accordingly. For example, we trust visual signals by day and auditory signals by night. We assign greater weight to what we see by day and to what we hear by night, because we have *learnt* to do so. Accordingly, we expect visual information to be more precise by day than by night. This is called 'expected precision'. Because we do not expect visual precision by night, we tolerate all sorts of blurry and fuzzy images without a second thought; but similar visual experiences by day would make us think there was something seriously wrong.

Likewise, Eve Periaqueduct's staff learnt that falling barometric pressure predicts increasing precipitation in winter, but not in summer. Therefore, they could treat a falling reading as 'noise' more confidently in the context of summer than in winter. They would *expect* barometric readings to be less precise in summer. If they proved not to be, they would adjust their expected levels of precision, which would in turn affect their posterior predictions. To take a more affective example: those Sussex sheep learnt to trust their prediction that people running towards them during the day cause no harm, but they are less confident of that prediction at night. They accordingly assign less expected precision to it at night. Perhaps, if they continue to encounter me in conditions of darkness and reliably come to no harm, they will adjust this precision value and accordingly alter their predictions about people running towards them at night. Contextual dependencies can be learnt, just like anything else.

It is clear from what I have said that this type of learning revolves around *fluctuating contexts*. Without predictive modelling of contextual dynamics, a self-organising system cannot survive for long in changing environments. The generative model must incorporate these dynamics. It must learn to predict degrees of precision. And the adjustment of precision values, like everything else in the predictive brain, must follow Friston's Law.

Precision is how the brain represents its degree of confidence in a given source of sensory evidence or in the predicted consequences of a given action. Precision values quantify expectations about variability. So, they are *representations of uncertainty*. How confident am I about this error signal in the present context? How much *weight* should I give to it, right now? Is 8/10 for *A* worth more or less than 8/10 for *B under present conditions*?

We have already seen the physiology of this: the midbrain decision triangle prioritises a need, then the forebrain's model of the self in the world generates an expected context in which that prioritised need will be met. There are two facets to this expected world: the actual content of the predictions and the level of confidence the system has in those predictions. Now that we know how these levels of confidence are quantified, we can slot them into our account of the physiology of arousal.

After the decision triangle has selected its currently salient need, which determines the affective state of the system, which in turn determines the expected context that is generated by the forebrain's long-term memory systems, the reticular activating system sets to work. The memory systems assign *baseline precision values* for the expected context and apply them throughout the predictive hierarchy. Then a cloud of neuromodulators billows through the forebrain, urging some channels to fire rapidly and discouraging others. These firing rates determine how much weight will be given to the current predictions and their attendant errors, which governs how 'loudly' the errors will be transmitted. In other words, the precision values determine *how confident* the system is about the outcomes it expects to follow from the course of action that now unfolds, over the various levels of the hierarchy. Then, once again, it hopes for the best but prepares for the worst.

Lots of surprising things can be predicted once you get used to them. Nevertheless, if hoping for the best and preparing for the worst is all we can do, this implies that some things *cannot* be predicted. And that's the second part of the story. It requires the system to adjust its confidence levels on the hoof – that is, to *modulate* arousal in the context of unfolding events, as they happen.

I said earlier that it is conceivable for an extremely complex set of model algorithms to evolve (no matter how unwieldy they become) which compute relative survival demands in all predictable situations, and to prioritise its action options on this basis, notwithstanding the 'combinatorial explosion'. But how does the organism choose between A and B when *uncertainty itself* becomes the primary determinant of action selection? This is what happens in novel situations, for example, which are far from rare in nature.

What physiologists call 'arousal modulation', computational scientists call 'precision weighting'.[5] They are the same thing. As you have just seen, a precise signal is nothing other than what I called a 'loud' signal in Chapter 6 – it is a *strong* signal. This implies that the modulation of confidence in an error signal must follow *deviations from its expected strength*. And these deviations must be minimised. As with all homeostatic error signals, it is 'good' (for us biological systems) when things turn out as expected, and 'bad' when uncertainty prevails.

As the action sequence unfolds, baseline confidence levels are *adjusted* upwards and downwards by the reticular activating system. (Think of Eve Periaqueduct's meter readers and adjusters.)

That is, the unfolding sensory-motor context is 'palpated' – and the system's confidence weightings are adjusted – on the basis of unfolding fluctuations in expected uncertainty. Alterations in arousal track the estimated reliability of the sampled prediction errors. In this way, fluctuating precision values estimate the changing reliability of the unfolding signals carrying the news. These values, in turn, determine everything else the system does, in accordance with Friston's Law.

All of this suggests that *precision optimisation is the statistical-mechanical basis of signal prioritisation in general* – that is, it is the critical output of everything we saw going on in the midbrain decision triangle and reticular activating system. Precision optimisation is how the multiple error signals converging on the PAG were prioritised in the first place, bringing the most salient need to affective awareness, leading to a series of unfolding choices in an expected context, guided by expected precisions. These must now be modulated on the basis of unexpected sensory events.

Perhaps this sounds rather abstract. I think it is, on the contrary, very true to everyday life. So much of our experience just is little pulses of feeling, as you notice things that aren't quite as you expected them to be, followed by cognitive castings around for ways to close the gap. You remember an email you need to send: it is only when your hand fails to detect the hard screen of your phone that you realise you were already reaching for it – but if it isn't right there beside you, where *did* you leave it? In the kitchen, where you were five minutes ago?

Or take this rather more affectively significant example: an encounter with a potential lover. You think you might win him or her over tonight. You picture the possible sequence of events and you set a plan of action. Then you hope for the best. You are not sure how things will turn out, but, based on your previous experience with this person, you rate your chances at about 7/10.

As the evening unfolds, what you focus on (what is salient) is quite different from what you would have focused on if you were having dinner with your brother. The emotional tone is different too. Every little sign suggesting that the person across the table from you is responding positively to your overtures elicits rushes of excitement: increasing confidence that your plan is working. Unexpectedly, your hoped-for lover yawns and looks at their watch. What does that mean? How much weight should you give to this development? You have a sinking feeling. Have you been misreading all the previous signs? You scrutinise every movement and gesture. The slightest further indication that your feelings are not reciprocated will prepare you for the worst, and you will initiate Plan B: salvaging your pride by pretending you are likewise indifferent. But then your eyes meet. Does that mean what you think it does? Yes it does! Next, your hand is being gently touched. Your heart quickens. It looks like you can stick to Plan A after all.

A prioritised need (in this case LUST) is the currently most salient source of uncertainty. Inferences about its causes become conscious as affect, because fluctuations in your confidence level concerning the possible actions required to meet this need must be modulated by feelings. The feelings tell you how well or badly you are doing. The unfolding context giving rise to the fluctuations must become conscious too, for the same reason. That is why I defined exteroceptive awareness of action and perception as *contextualised affect*. Now we have a formal, mechanistic grasp on what this means. All of it is just felt uncertainty.

It is of capital importance to note that the statement 'the unfolding context giving rise to the fluctuations must become conscious too' explains why experience has dual aspects. It is not merely a matter of 'I feel like this' but rather 'I feel like this *about that*'. The 'about that' must be felt, too, using a common currency (applied uncertainty) – because context is the main source of uncertainty over free energy. The economics of free-energy minimisation demands a common currency.

These facts reveal that consciousness is not merely a subjective perspective upon the 'real' dynamics of self-organising systems; it is a function with definite causal powers of its own. The *feeling* of a need (as opposed to the mere existence of a need) makes a big difference to what the subject of that need will do next. Affects literally drive what an animal does from moment to moment in conditions of uncertainty. The whole purpose of exteroceptive perceptions is that they are felt *in relation to* the affectively driven actions they contextualise.

This is the central function of things like attention. Attentional focus works like affect selection, but it is applied to the outside world. Our need to reduce uncertainty governs our gaze, for example, so that visual saccades track the regions of a scene where more precise information is likely to be found.[6] Simply put, relatively strong signals attract attention: they are assigned higher precision.

That is how saliency works. 'Salient' features of the world are features that, when sampled, minimise uncertainty concerning the system's currently prioritised hypothesis: they are the ones that, when things unfold as expected, maximise our confidence in the hypothesis. Active agents are thus driven to sample the world so as to (attempt to) confirm their own hypotheses.[7]

Since these are, in the final analyses, hypotheses about how to meet our needs, this means that each species is driven to select its own perceptual world. The perceptual orientation of each species is dictated by the things that matter to it. Accordingly, humans, sharks and bats live in different (subjective) worlds. You perceive objects and events only when you notice them, and different ones are salient to each species. You can only see what you sample.[8] The Chilean biologist Francisco Varela put it nicely: 'the species brings forth and specifies its own domain of problems'.[9]

This implies that precision cannot be determined passively; we cannot just wait and see which signals are strong without any expectations either way. It must be inferred and then *assigned* by the generative model. Attention – which has everything to do with precision – can accordingly both be 'grabbed' and 'directed'.[10] For example, we actively reduce precision on sensory errors almost to zero when we go to sleep, but a sufficiently surprising event will still wake us up. In other situations, we might actively amplify it, such as when we apply ourselves to a dense text because we suspect it contains something important.

In the (2018) article that set out our theory of consciousness, Friston and I altered some of the conventional symbols used in equations that calculate free energy (see the endnotes to Chapter 7). We replaced them to acknowledge the fact that we are following in the footsteps of Sigmund Freud, who attempted in 1895 to 'furnish a psychology that shall be a natural science: that is, to represent psychical processes as quantitatively deter-minate states of specifiable material particles, thus making those processes perspicuous and free from contradiction'.[11]

Those were the opening lines of his 'Project for a scientific psychology'. There, Freud used the symbols φ, ψ, ω and M to denote four hypothetical systems of neurons, responsible for perception, memory, consciousness and action, respectively; and he used the symbol Q to denote external stimuli. Friston and I followed this usage to denote the equivalent vectors within a self-evidencing system:[12]

$Q\eta$ = external states, as modelled by the system's internal states

φ = sensory states

M = active states

ψ = predictions

ω = precisions

In addition, we used:

e = prediction errors (based on φ and its prediction ψ)

F = Friston free energy (based on e and precision ω).[13]

Notice that none of these quantities measures external states directly, since they are *hidden* from a self-organising system. This means, in everything that follows, we have a self-contained, autonomous description of mental dynamics in terms of the system's own internal $(Q\eta, \omega)$ and Markov blanket (φ, M) states. Equipped with these terms, we can now formalise a self-evidencing system's dynamics in relation to precision optimisation.

I will start with two equations that define Friston free energy in terms of the quantities I've introduced. The first equation says that 'free energy is (approximately) the negative logarithm of the probability of encountering some actively authored sensory states'.[14] The second one says that 'the expected free energy decreases in (approximate) proportion to negative log precision'.[15] Remember that the whole point of the self-evidencing system's dynamics is to minimise free energy.

With these relationships in place, it becomes evident that there are in fact *three* ways for a self-evidencing system to reduce prediction error and thereby minimise free energy – not just the two obvious ones I described previously:

(1) It can *act* (i.e. change M) to alter sensations (φ) so that they match the system's predictions. This is action.
(2) It can *change its representation of the world* ($Q\eta$) to produce a better prediction (ψ). This is perception.

And now, in addition:

(3) It can *adjust precision* (ω) to optimally match the amplitude of the incoming prediction errors (e).

This, I submit, is consciousness.[16]

It is this final optimisation process, the optimisation of the system's *confidence*, as described in the text above, that Friston and I associate with the evaluation of free energy that underpins felt

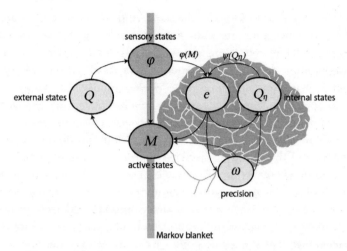

Figure 17 The dynamics of a self-evidencing system equipped with precision optimisation. The symbols are explained verbally in the text. (*Q* denotes external reality itself, which is hidden from the system and therefore does not appear in the equations.)

experience. The equations formalising these dynamics are given in the endnotes.[17] Since the third equation is the crucial one, I will state it in words: 'The rate of change of precision (ω) over time depends on how much free energy (*F*) changes when you change precision. This means that precision will look as if it is trying to minimise free energy.[18] The rate of this free-energy-minimisation process is the difference between the variance (the inverse precision) and the sum of squared prediction errors (*e.e*).'[19] In more basic terms, the third equation quantifies how *ongoing adjustment of precision implements Friston's Law*, alongside action and perception. In short, it quantifies how *consciousness contributes to action, perception and model updating* – and thereby to minimising free energy.

Figure 17 represents these dynamics visually.

Conceptually, precision is a key determinant of free-energy

minimisation and therefore the activation of prediction errors. Precision determines which prediction errors are selected – and therefore, ultimately, how we represent the world and our actions upon it. If precision is 'arousal' (which it is), this explains, formally and mechanistically, why confidence optimisation is always and only an endogenous process. Consciousness *must* come from within.[20]

The process I have introduced here manifests in many guises. Let's consider it once more in biological (i.e. physiological and psychological) terms. In the *exteroceptive* domain, it manifests as attention and attenuation, associated with the increase and decrease of sensory precision.[21] In the *proprioceptive* domain, it corresponds to the precision of motoric affordances (possible uses of objects), of the sort associated with goal selection and real-isation.[22] In the *interoceptive* domain, it literally determines 'gut feelings', i.e. the best explanation for interoceptive signals that have been enabled or aroused.[23] However, it is very important to note that all of these things (exteroception, proprioception and interoception) can occur without consciousness; consciousness is the *feeling* of these things.

To sum up: the task of precision is to *arouse* representations (and expectations). In the absence of precision, prediction errors will fail to induce any perceptual synthesis or motivated behav-iour. In other words, without precision, prediction errors would be sequestered at the point of their formation. That is what hap-pened in Oliver Sacks's akinetic-mute patients, for example.

This formulation of precision implicates the neuromod-ulatory mechanisms described in Chapter 6. These are the mechanisms that generate altered states of consciousness[24] and dreams[25] and they are targeted by consciousness-altering drugs (i.e. psychotropics and psychedelics).[26] This formulation of preci-sion also lends some validity to neuromodulatory versions of the 'global workspace' theory of consciousness (discussed briefly in Chapter 4).[27]

Unsurprisingly, the role of precision in psychopathology is

an important theme in the burgeoning field of computational psychiatry.[28] Recall the earlier case of Mr S, in which we saw what happens when error signals are assigned too little weight. We can see from Figure 17 how his decision triangle and therefore reticular activating system (ω) gave too much weight to his predictive model ($Q\eta$) and too little to his prediction errors (e). As you saw, this had everything to do with his feelings.

I know it seems odd to talk about consciousness in mechanistic terms like these. That is because I have been describing the laws underlying phenomenal experience rather than phenomenal experience itself. By setting out these laws, I have been trying to show that consciousness is part of nature; that it does not exist in some parallel universe; it is not something beyond the reach of science.

Now I am going to ask you to cross a Rubicon with me.

This chapter addresses the questions why and how consciousness arises, not biologically but *formally and mechanistically*. More specifically, it asks: (1) Why and how does the biological need-prioritisation process I described before relate to the *laws* of free-energy minimisation? (2) Why and how does it happen that this mechanistic process *causes* some self-organising systems to feel like something? The question I have been addressing over the last few pages, from the perspective of statistical physics, is the first one. So, now I must explain in a straightforward way why and how the statistical-mechanical dynamics I have described generate felt experiences. How can a mere information-processing system have feelings? To explain this, I must ask you to take a leap that many natural scientists are reluctant to take – much to the detriment of science, and especially of mental science. It is to consider the mechanistic dynamics I have described *from the viewpoint of the system*. I am asking you to replace the third-person, objective perspective we have taken on the dynamics so far in this chapter with a first-person one: with the *subjective* perspective of the self-evidencing system itself. I am asking you to adopt the system's point of view, to *empathise* with it.[29]

This leap is justified by two facts I have already explained. The first is that felt experience can be registered only from the subjective perspective. To rule the subjective perspective out of court is therefore to exclude from science the most essential feature of the mind. This is what the behaviourists did, setting the stage for neuroscience's half-century of failure to grapple meaningfully with consciousness. The second fact is this: I have shown, in formal and mechanistic terms, how the subjectivity of self-evidencing systems comes into being. Therefore, taking the subjective perspective of a self-evidencing system is justified precisely by the fact that it possesses selfhood.

To be clear about what I am saying: free energy and its constituent precisions are only experienceable within a system when it is subjectively conceived, from the viewpoint of the system; experiences cannot be observed *as experiences* from without, objectively.

I have also explained in causal, law-following terms why and how the selfhood of such systems is *intentional*. Self-organising systems with the dynamics I described have an aim and a purpose, namely to survive. This means that they possess a value system: the same value system that underwrites all life.

The intentionality of dynamical self-organising systems *obliges* them to ask questions about their own states in relation to the entropic perturbations that surround them. This is what makes them 'self-evidencing' systems. They *must* always ask, 'What will happen to my free energy if I do that?' Moreover, complex self-evidencing systems (like us vertebrates) must ask this question in relation to *multiple* categorical variables; so, the answers – our vital statistics – must be both quantified *and qualified*. Finally, they must modulate their level of *confidence* in the answers they receive.

What I am describing here in abstract, technical terms is nothing too complicated. You know it from your personal experience. What you experience all the time is fluctuating pulses of feeling in response to your movement through the world, as you check whether everything is as you expected to find it – and as you try to close the gap, somehow, when it isn't. Isn't that basically how experience is for you?

Combining all these facts about the subjectivity and intentionality of complex self-evidencing systems, we arrive at the following conclusion. The equipment-evoked responses (in Wheeler's sense) which flow subjectively from the types of question that systems like us are obliged to ask must possess existential value and multiple qualities. Our confidence in these fluctuating responses – the 'phenomena' – that we register must be subjective, valenced and qualified.

And that *just is* what it is like to experience consciously. The equipment-evoked responses, in the case of us vertebrates at least (and no doubt some other organisms too), are felt.

To help you across the Rubicon, please remember that feelings *evolved*. The dawn of consciousness gave rise to very simple phenomena, like feeling too hot. Increasingly precise sensory evidence that predicts the demise of an overheating self-evidencing system just feels like 'too-hot-ness' to the system. It took aeons for such elemental forms of affect to became elaborated, over a deep predictive hierarchy, and ultimately to produce Merker's 'fully articulated, panoramic, three-dimensional world composed of shaped solid objects: the world of our familiar phenomenal experience'.

In such a world, we feel what it is like to be a system with the dynamics I have described. Feelings are fluctuating, existentially valued, subjective states with differentiated qualities and degrees of confidence. This is the stuff of consciousness. Now we see why it has to be that way.

10

Back to the Cortex

As we have seen repeatedly throughout our journey, the cortical fallacy has a lot to answer for. Had the pioneers of behavioural neuroscience not been so impressed by the large expanse of our cortex or been so blinded by the philosophical idea that mental life arises from associating memory images, we might have discovered the real source of consciousness a good deal earlier. It is a tantalising irony of the history of mental science that Freud possessed so many pieces of the puzzle more than a century ago. The clues, both neurological and psychological, were staring him in the face. But, when it came to consciousness, even he fell prey to our collective fixation with the cortex – an obsession whose cost, in case we forget, may be measured in more than just wasted time.

All this is true. Yet the cortex clearly has a massive role; our everyday experience is intimately bound up with the dynamics of cortical processing. In this chapter, then, let's return to this misunderstood upper storey of the brain to see what it adds to our account of consciousness. As we shall learn, many of the most commonplace features of everyday experience derive their character from what the cortex does – but not in the ways that we thought previously.

The most obvious sense in which this is so is that the world as we experience it is literally generated from cortical representations. Within the predictive coding framework, odd as it seems, what we perceive is a *virtual reality* constructed from the mind's own building materials.

This is a radical notion when compared to the common-sense view, but the idea that perceptual experience is self-generated is widely accepted in contemporary neuroscience. Take for example what Semir Zeki said about colour vision as long ago as 1993. He wrote that colour is '*a property of the brain*, a property with which it invests the surfaces outside, an interpretation it gives to certain physical properties of objects'.[1] He elaborated:

> Suppose one looks at a patch illuminated with long-wave light in isolation [...] The patch produces a high lightness record for light of any waveband, since the only comparison that the brain can undertake in these conditions is a comparison of the light reflected from the illuminated patch and the dark surround. The long-wave light thus produces a high lightness while the middle- and short-wave light, being absent, produce no lightness at all. The nervous system thus assigns the colour red to the patch.[2]

Note Zeki's choice of words: the brain *assigns* redness to the world, after asking questions about the relative intensities and wavelengths of light. It paints the world by numbers. The same applies to the phenomenal properties that characterise our other modalities of perception: sounds, tastes, somatic sensations and smells. The brain assigns these qualities to the world.

I suspect that many readers find it hard to believe that what they are seeing right now is not simply what is 'there'. I can imagine you asking: 'Where else does my perception of these words on the page come from?' It might help if I point out that what you are seeing right now bears little resemblance to the sensory inputs you are receiving. Those inputs start out as light waves impacting on your retinae. The photosensitive cells there (called rods and cones) respond to the light waves by generating nerve impulses. These impulses – not the light waves themselves – are then propagated along your optic nerves to the cortex, in the form of spike trains (see Figure 11). Why do you experience these trains – 0011111101101 – as moving images out there in the world?

The neurons in the lateral geniculate body and occipital 'projection' cortex that respond to the retinal impulses are arranged topologically, thereby creating the possibility of image-making through *mapping* the retinal surfaces (see Figure 6); but this does not take account of the fact that there are no rods and cones near the centre of your retinae, where the optic nerve emerges. By rights, therefore, you should see a black hole near the middle of your visual field. How does the hole disappear? The answer is that you *infer* what belongs in the 'blind spot', from context and from memory, and then you fill it in.[3]

I should have said there is a black hole in your visual fields (plural) because, let's not forget, you have two of them. This raises another question: why don't you see two images? I am not referring to the fact that you have two eyes; it is easy to imagine how two almost identical maps can be superimposed, but that is not what happens.[4] What happens is that the cells in the left halves of both retinae project to your right occipital lobe, and those in the right halves project to the left lobe. This means that what you actually have in your visual cortex are two different representations of the retinal surfaces (one of the left half of this book and the other of the right)[5] with an anatomical chasm between them: the longitudinal fissure which divides the cerebral hemispheres. How do the two fields become the unified image you are seeing? (It is true that they are coordinated by axons across the corpus callosum; but people in whom the corpus callosum is missing see a single image, too.)[6] In addition, we must take account of the fact that your visual fields – as they are represented in the occipital lobes – are upside down and back to front in relation to the images you see. Furthermore, your eyes dart all over the place, about three times per second, not to mention the constant movement of your head. How come you perceive a stable, correctly oriented visual scene?

My point – that there is little resemblance between what you see and the sensory inputs arriving at your cortex – is vividly illustrated by neurological patients in whom there is damage to the mechanisms by which we normally convert what arrives there into what we see. I reported one such case many years ago: a twelve-year-old boy (WB) with bilateral abscesses in his frontal lobes, caused by rampant sinusitis.[7] He periodically saw the world rotated through 180°. The symptoms and signs in this patient were identical to those described in twenty-one previous case reports I found scattered through the world literature, since 1805, lending credibility to his subjective description (see Figure 18, for some objective evidence).

I currently have an even more interesting patient who is being researched by my doctoral student Aimee Dollman. I cannot report all the details, because the case is still under investigation and the findings are not yet published. The patient is a highly intelligent young woman with cortical dysgenesis (abnormal anatomy) of the occipital lobes, who represents the world in almost exactly the way I just said we do *not* experience it: as it is arranged anatomically in the visual cortex. This occurs especially when she uses her visual memory (i.e. her predictive model). She sees two separate fields, upside down and back to front (not always in unison with each other). Her visual model of the world *fails to make the usual corrective inferences* by which we orientate and integrate the visual fields. It therefore makes erroneous predictions and her visual experience does not square with that of her other sensory modalities. Accordingly, she sometimes gets muddled about the direction that her body is moving in (especially when travelling in trains and aeroplanes), and she makes gross errors when navigating the environment. In addition, she has difficulty inferring invariant objects from fleeting visual data (such as the spelling of words in her mind's eye when she has to abstract them from variable scripts and fonts, or the identity of faces abstracted under fluctuating lighting conditions and different viewing angles).

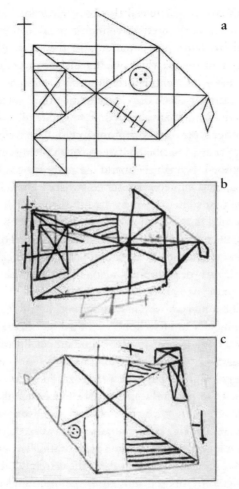

Figure 18 (a) = the Rey Complex Figure as shown to Patient WB; (b) = the same figure as copied by him; (c) = the figure as drawn by him from memory. These drawings provide objective evidence of the inversion of WB's predictive model of the world. The Rey Complex Figure is difficult to reproduce at the best of times; why would this seriously ill patient compound the difficulty by drawing it upside down?

In other words, her visual cortex receives the sort of information that we all do, but she cannot generalise from the noisy sensory signals and automatically infer the stable objects they stand for (e.g. recognise a familiar face). She has suffered these anomalies all her life, and has developed elaborate ways of compensating for them. Since her visual 'association' cortex does not *automatically* integrate the two visual fields and rotate the scene, she adjusts her representation of the world by making *deliberate* inferences. For example, when I asked her to identify the location of a well-known city on an unmarked map, she said: 'Should I show you where I *sense* it is located or where I *know* it is?' When she 'senses' that something is located in the west, she 'knows' it must be located in the east.

Rather than appealing to such rare neuropsychological disorders, however, let me illustrate the self-generated nature of perception by way of the phenomenon that is conventionally used for this purpose, namely *binocular rivalry*.

This phenomenon was first described in 1593; and it featured prominently in Helmholtz's seminal work on the topic of unconscious inference.[8] It involves simultaneous presentation of different pictures to each eye, using a mirror stereoscope. Let's say the left eye is presented with a face and the right one a house. Under these artificial conditions, visual experience unfolds in a 'bistable' manner, whereby you do not see a superimposed blend of the two images but rather an *alternation* between them. You see a house then a face then a house then a face, rather than a combined house-face. This clearly demonstrates the distinction between the objective signal that is transmitted to the brain and the subjective percept that is generated by it. Helmholtz concluded: 'In such cases the *interpretation* [of the visual signal] vacillates such that the observer has different experiences, one after another, for the unchanging retinal image.'[9] As with colour vision, therefore, what you experience is an *inference* about the sensory input, not the input itself.

Broadly similar things happen in everyday life, such as when I 'saw' my British friend Teresa at Cape Town airport. All these illusions show that what you perceive is generated largely by your *expectations*. In Bayesian terms, binocular rivalry is taken to show that if the *prior hypothesis* that best fits the sensory data (the high likelihood that you are seeing a house-face) does not square with your *background knowledge* (the low probability that house-faces exist) then the hypothesis is rejected. The inference that you're seeing a house trumps the one that you're seeing a house-face; so, a house is what you experience. But when you test this posterior hypothesis (as a new prior) it fits only half the sensory evidence. Your background knowledge dictates that whole faces are just as probable as whole houses are. So, you change your mind and infer that you must be seeing a face; which is what you then experience. But when you test this new hypothesis, it again does not account for half the sensory evidence. And so on ...

The Bayesian interpretation of binocular rivalry is widely accepted. What clinches it for me is the fact that when people are shown two images which in combined form *do* have a high prior probability, they perceive a blended image. For example, dichoptically presented images of a canary and a cage are readily seen as a canary in a cage.

What you perceive is not the same thing as the input that arrives from your senses. *What you perceive is an inference.* And the materials from which that inference is derived are for the most part your cortical predictive model derived from past (i.e. expected) experiences.[10]

That tells us something about what cortex brings to consciousness. But what about the other way around? What does consciousness do for the cortex?

What I am going to tell you now is obvious, yet nobody else seems to be saying it.[11] It is this: cognitive consciousness is generated by a recently discovered neural mechanism called 'memory reconsolidation'. Like so much else, the idea has its origins in something Freud wrote. He wrote that *'consciousness arises instead of a memory trace'*.[12] What he had in mind was slightly different from what I am going to tell you, because Freud – like all the neurologists of his day – was trapped in the cortical fallacy.[13] Nonetheless, it is again rather astonishing how close he got to the truth.

We have seen that affects make demands upon the mind, and that cognition performs the work so demanded. To be more precise, *conscious* cognition performs the work; for once it has been performed, and confidence in the (prioritised) belief that had become uncertain has been restored, the generative model resumes its automatic mode of operation, below the threshold of awareness.[14] Here once again is the mechanism of *learning from experience* that I have described repeatedly. This is the whole point of consciousness in cognition. You arrive at a situation in which you aren't sure what to do. Consciousness comes to the rescue: you feel your way through the scenario, noting the voluntary actions that work for you. Then, gradually, the successful lessons become automatised and consciousness is no longer needed.[15]

I want to emphasise that the cognitive work I have just described *slows down* the otherwise automatic business of acting in the world. This is the essential difference between voluntary and involuntary action, conscious versus unconscious cognition, felt drive versus autonomic reflex; the voluntary type is *less certain* and therefore requires more time. This process, which delays automatised action tendencies and enables them to be held in mind (in short-term memory), is aptly called 'working memory'. Working memory is literally the holding in mind of feeling – stabilised affect transformed into cognitive work. As I just said: if affect is demand made upon the mind for work, then conscious cognition is the work itself. Thus, affect both accompanies and *becomes* cognition. The 'work' in question entails inhibition of automatic action tendencies and the *stabilising* of intentionality while the system feels its way through unpredicted problems. This bestows considerable adaptive advantages, as it facilitates viable solutions to the many real-world problems that our generative models cannot (yet) predict. This stabilising process is *the* function of cortex.[16] Cortex specialises in uncertainty.

What all of this implies is that *the conscious state is undesirable* from the viewpoint of a self-organising system. Look back at the dial diagram in Figure 12: the outer arrow represents increasing demand for work (negative affect) and the inner one represents decreasing demand (positive affect), but the ideal state is the settling point which represents *no demand at all*. In Chapters 7–9, I placed these matters on a formal mechanistic footing. Free-energy minimisation is the ideal state of living systems, which means that minimal surprisal is the ideal. This means, simply, that minimal *need* is the ideal. Affect is nothing other than the announcement of salient need. This should mean that feeling is a good thing because it enables us biological systems to resolve our needs and thereby avoid destruction. But the *ideal* state is surely one in which all our needs are met automatically – even before they are felt – i.e. where there is *no uncertainty*. In that theoretical ideal state, in which our needs are met automatically, we feel nothing. (This is how most of our bodily needs are met: they are regulated autonomically.) I say 'theoretical ideal' because, in respect of many of our needs, especially the emotional ones, we never get there. The SEEKING drive alone ensures that. This is the bad news.

The good news is that the error signals with the highest precision values have the greatest influence over the generative model. Because they demand the most change – because they declare that you are doing something *wrong* – they represent the greatest opportunities for learning, so that you will not find yourself in the same pickle next time. This is no different from 'learning from your mistakes', as your parents and teachers were always telling you to do. Now you know why: they wanted to spare you from oblivion. But the surprising upshot remains that *consciousness is undesirable* in cognition. What we are all aspiring to, therefore, is not pleasure (decreasing need) but zombiedom (no need). No need implies perfect predictions, which means no errors, and therefore no call to increase the precision on incoming signals, and therefore no feeling. Peace at last.

Freud's aphorism, 'consciousness arises instead of a memory trace', should make more sense now. It means that *consciousness arises when automatic behaviour leads to error*, in other words, when the memory trace (a prediction) producing a behaviour does not have the expected outcome. This means that the prediction in question must be *updated* to accommodate the error. Cortical consciousness may therefore be described as '*predictive work in progress*'. A memory trace that is conscious is in the process of being updated. It is no longer a memory trace. Hence: consciousness arises instead of a memory trace.[17]

In the last years of the twentieth century we understood the neurophysiology of this process in terms of 'reward prediction error' (see Figure 19).[18] With the start of the twenty-first century, we have obtained a tighter grip on the physiology of memory updating, under the heading of 'reconsolidation'.

The basic facts first came to light with the discovery in the 1960s that a fear memory can be eliminated using electroconvulsive shock therapy (ECT) – but only if the shock is administered immediately after retrieval of the memory.[19] This suggested that ECT interferes with a process which returns manifest (activated) fear memories to their latent state: if the shock is administered while a fear memory is activated, then the memory is wiped out. It is literally *no longer a trace* while you are remembering it. The activated state of the memory makes the long-term trace *labile* once more: it makes it no longer a memory.

This was confirmed when reconsolidation was rediscovered and named as such in 2000.[20] If protein synthesis inhibitors are administered while a long-term trace is activated, the trace disappears. (Protein synthesis inhibitors prevent *new* long-term traces

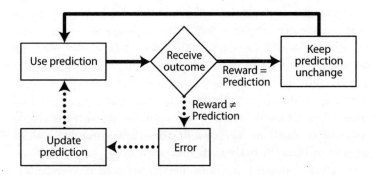

Figure 19 Scheme of learning by reward prediction error. The sequence begins with the box 'Use prediction', which leads to two possible outcomes, to the right of the box. Either a prediction error exists (when the outcome differs from the prediction), which leads to the box 'Update prediction'. Alternatively, no error exists (when the outcome matches the prediction), which leads to the box 'Keep prediction unchanged'; then the behaviour remains unchanged. The 'Update prediction' stage corresponds to what we now call reconsolidation.

from being formed.) This applies to other kinds of memory too, not only to fear memory. Long-term memories, in general, become unstable when they are in the activated state. That is how they are updated (and then consolidated afresh – i.e. reconsolidated).

An enormous amount of research into reconsolidation has been done over the past two decades, showing that it occurs not only in humans and rodents but also in chickens, fish, crabs, freshwater snails and honeybees. These studies have also demonstrated an analogue of reconsolidation in pain-processing pathways of the spinal cord, suggesting a very basic role for it in the central nervous system.

Long-term memory, unlike short-term, is dependent upon the synthesis of new proteins, which is triggered after substantial and repetitive synaptic transmission, which in turn is modulated by the reticular activating system – that is, by arousal.[21] Hence the famous adage known as Hebb's Law: 'Neurons that fire together wire together'.[22] An 'activated' memory is an aroused memory; and an aroused memory is a memory no longer – it is in a state of uncertainty. All I am trying to convey here is that cognitive consciousness boils down to a rendering labile of cortical memory traces, and that this lability is *a product of arousal*. We keep arriving from different directions at the same insight: cortical processes are fundamentally unconscious things (they are simply algorithms, if left to their own devices). Consciousness – all of it – comes from the brainstem.[23]

But still, we seem to strive for a kind of zombiedom. The ideal form of cognition is automaticity, and so the sooner we can get rid of consciousness, the better. How, then, does cortical cognition become *unconscious?*

Let us consider the simplest possible example. What do you think would happen if I projected a vertical line in front of your eyes in such a way as to discount the constant eye movements which normally ensure that everything you look at is updated about three times per second? This updating happens at the very periphery of the predictive hierarchy, where the largest degrees of freedom are found. The frenetic rate of model updating required in such unpredictable circumstances makes the word 'prediction' almost meaningless. That is why so much cognitive work occurs at the sensory-motor periphery.

So, what would happen if one could immobilise a visual stimulus in such a way that we saw it and only it absolutely monotonously? The unsettling answer is that *it vanishes from consciousness*. Even though the stimulus is still there, it disappears from visual awareness within a few seconds. This was demonstrated in the 1950s by Lorrin Riggs in America and Robert William Ditchburn in England. Similar effects have since been observed for other sensory modalities.[24] The reason the stimulus fades from awareness should be obvious to you by now: it becomes 100 per cent predictable and therefore carries no information. The prediction achieves total precision and the corresponding error value drops to nothing. This, as I have said, is the brain's homeostatic ideal.

It is difficult to stabilise stimuli relative to eye and head movements, so you can't do the Riggs–Ditchburn experiment at home. You can however Google the 'lilac chaser' illusion, which shows another type of visual fading.[25] Once you have clicked on the image, hold your face about 20 cm from the computer screen and focus on the target in the middle of the rotating circles. Then notice what happens to the circles. They disappear – because your visual brain infers that they are noise and assigns them less precision relative to the grey background, fading them out.

This web site also shows two other visual illusions. In addition to the fading, you will see a lilac circle turn into a green one which is not really there – neither (1) really present nor (2) really green – spinning around the central target. The web site explains the mechanisms at work. They basically boil down to precision weighting. All these phenomena underscore the point that what you see is generated by your brain rather than received from outside.[26]

Let's link this with the section in Chapter 5 about the updating of innate motor priors: reflexes and instincts. These innate predictions serve us well, but they cannot possibly do justice to the complexities of the world, so they must be supplemented through learning from experience. Learning requires consciousness, as we gradually improve confidence in our newly acquired predictions. But the ideal of all learning is to *automatise* these acquired predictions, too, to make them behave like reflexes and instincts. We aspire to forge new predictions that are at least as reliable and generalisable as the old ones. To the extent that we can achieve this, acquired predictions become automatised through systems *consolidation*. Consolidation is, in this sense, the opposite of reconsolidation, which undoes consolidated traces, literally dissolving the proteins that 'wired' them.

This happens right down to the level of the non-declarative memory systems. The goal of learning from experience is to shift as many long-term memories as possible from the declarative to the non-declarative state, for the reason that 'declarative' means 'capable of returning to consciousness'. So, when I said that zombiedom is the ideal of cognition, I meant that ever deeper consolidation is the ideal of learning. Non-declarative memory is the *most reliable* form of memory. It entails the least amount of work. It minimises complexity and is the most generalisable (see Chapter 8). Accordingly, it is the quickest to execute; it involves the least uncertainty and therefore the least delay.

There are, of course, some complexities here. Firstly, I do not want to give the impression that all consolidation proceeds from declarative into non-declarative memory systems. Many long-term predictions are consolidated *directly* into non-declarative memory, and most types of learning occur over both systems simultaneously. Second, there are multiple types of non-declarative memory, and they do not all work in the same way. For example, 'procedural' learning occurs through brute repetition; hence we say that skills and habits such as riding a bicycle are 'hard to learn and hard to forget'. Yet some varieties of non-declarative emotional response, which are equally hard to forget, are acquired by single-exposure learning: fear conditioning, for example. (Other types of emotional learning are slower; attachment bonding, for instance, takes about six months.) Hard-to-forget-ness is the cardinal feature of non-declarative memory; but consolidation entails very different processes in the multiple memory systems. Finally, non-declarative memories are only 'unconscious' in the *cognitive* sense. When an acquired emotional response is triggered you do feel something; you just don't know what the feeling is about – that is, where it came from (see Chapter 3).

The most important fact about non-declarative memory is that it is non-declarative. It generates procedural *responses*, whereas declarative memory generates experienced *images*.[27] This coincides with an anatomical distinction: declarative memories are cortical while non-declarative ones are subcortical.[28] Subcortical memory traces cannot be retrieved in the form of images for the reason that they do not consist in cortical mappings of the sensory-motor end organs. They entail simpler stereotypes, of the kind that I described earlier in relation to the learnt behaviours of hydranencephalic children and decorticate animals.[29] Such things cannot be brought to mind; they are not 'thinkable'.

The cortical memory systems, by contrast, are always ready to revive the predictive scenarios they represent – literally to re-experience them. In other words, declarative memory readily returns long-term traces to the short-term state of conscious working memory. It does this not only to update the cortical predictions but also to guide action in conditions of uncertainty.

Subcortical memory traces are more reliable than cortical ones – their high precision values are less likely to change – because they are optimised for simplicity rather than accuracy. This makes them more generalisable. But it comes at a price: the less complex models are less accurate when the context varies.[30] The relative complexity of cortical predictions, on the other hand, coincides with greater plasticity. In a word, the cortex specialises in *contexts*; it restores model accuracy in unpredictable situations.[31] A trade-off is inevitable. The more potential for conscious experience, the less automaticity, which means more plasticity but also more cognitive work. That costs energy, and it generates feelings, so the brain does as little of it as it can get away with. Even to the point of fading out a stimulus that is right before your eyes.

And yet, a lot of what goes on in our heads seems hard to reconcile with this ideal of informational and thermodynamic efficiency. Along with feelings and perceptions, the other thing that our consciousness most conspicuously contains is *thoughts*. These are obviously cortical in origin – but what are they? And why do they often seem so idle?

The theory of cognition that I have been outlining here revolves around the capacity of our memory systems to generate a virtual world.[32] Each level of the predictive hierarchy – including each level of cortical processing – is capable of generating a version of the data that it expects to receive from the level beyond it. This means that perception is not fundamentally different from imagination: from the subjective point of view, there is little difference between the worlds you experience in your dreams and the one outside your window.[33] Your brain can conjure up phantasmal realities on demand. It is presumably doing so right now, as you read these words and think about what they mean. Even as you let your mind wander …

The existence of mind-wandering might seem like a challenge to the free energy theory of consciousness. I have said that we become conscious of only strong, prioritised error signals – *salient* ones – which we must respond to if we are to keep our biological parameters within viable bounds. Yet our thoughts often seem random and inconsequential. In some cases, we may even find our own internal monologues intrusive or distracting: unhelpful in the circumstances. How can that minimise free energy?

The aim of mind-wandering, odd as it might sound, is to improve the *efficiency* of your generative model. As ordained by the Free Energy Principle, a model is only efficient if it uses the minimum resources necessary to perform the work of self-organisation. That boils down to finding the simplest model that successfully predicts sensory samples of the world. (Remember Ockham's razor.)

The simplest model does not arise naturally from voluntary action. Voluntary action is a haphazard process. Simplicity is therefore increased by *pruning* redundant synaptic connections that formed while we were learning from experience. The reason for cutting them back is to avoid 'overfitting' our models to noisy data, needlessly preserving eccentric and weak correlations. The shears are the by now familiar mechanisms of memory consolidation and reconsolidation: by activating memories, we can strengthen, alter and even erase them.

Mind-wandering is one means by which this is achieved. It involves spontaneous forebrain activity (also known as the 'resting state' or 'default mode'), which occurs in the absence of any specific external stimulus. This kind of activity goes on much of the time, in the background, through an 'imaginative exploration of our own mental space'.[34] There is a good deal of overlap between this form of thinking and dreaming,[35] which seems to occur in all creatures equipped with a cortex: any animal with the capacity to generate images of itself acting in the world can also meander through endless simulated worlds as its circumstances permit.[36] Meandering, you might recall, is tightly bound up with the SEEKING drive, which continues with its demands as we sleep. It should be obvious why default mode activity is safer at night, when we do not have to contend with external events.

All of this accounts for the peculiar fact that our acquired models of the world are never quite stable – not even while we are sleeping. Even in the absence of compelling sensory inputs, structured neural activity continues, yielding ongoing exploration and testing of the generative model. These explorations, Andy Clark suspects, might result in elegant new responses to problems that have been occupying our waking attention – answers that are often simpler, in Ockham's sense, i.e. *more efficient* than our previous best attempts: 'Might all this be at least part of the solution to deep and abiding puzzles concerning the origins of new ideas and creative problem-solving?'[37]

Consciousness plays the same part in this self-generated process as it does in perception and learning from experience. What all conscious cognitive processes have in common is that they entail the necessary mental work of reconsolidation – the returning of consolidated predictions to states of uncertainty. That is why dreams (which are a form of problem-solving) are conscious.[38]

But there are other kinds of thought beside mind-wandering. Let's look at a second type: *deliberate* imagining. If any cognitive process deserves the antagonist role in the traditional opposition between thought and deed, it is this one. Here is thinking *instead of* action, inhibiting our motor impulses while the system feels its way through problems in imagination. When we engage in this type of thinking, we adjust our precisions so that motor prediction errors are suppressed.[39] After all, the point of deliberative thinking is to imagine doing things in order to gauge *in advance* the probable consequences of actually doing them. (I used the example of hitting my headmaster in Chapter 5.)

How do we imagine the future? In much the same way as we remember the past – which turns out, far more often than we like to admit, to be an imagined past. Episodic memory is a *constructed* process in which current goals and contexts contribute greatly to what is recalled. Thus, the past is selectively and tendentiously relived in relation to current demands for predictive work. Once again, the case of Mr S – the man with the 'missing memory cartridge' – provides an excellent example, precisely because the mechanisms underpinning normal remembering were so exaggerated by his pathology. His episodic memory was alarmingly self-serving.

There is, of course, no such thing as a memory cartridge. Instead, the neural systems involved in mental time travel pivot on the *hippocampus*, which is crucial for injecting the quality of perspectival 'mine-ness' into normally unconscious cortical memory processes.[40] Contemporary research on episodic memory reveals that the hippocampus is in fact just as involved in imagining the future as it is in reliving the past.[41] David Ingvar accordingly speaks of 'remembering the future'[42] and Daniel Schacter conceptualises the hippocampus – together with the other brain structures responsible for episodic memory – as supporting a 'constructive episodic simulation' of the future, which entails 'flexible recombination of details from past events into novel scenarios'. In

Schacter's view, the episodic memory system acquires its adaptive value more from its capacity to imagine the future than to remember the past. The brain, he concludes, is 'a fundamentally *prospective* organ that is designed to use information from the past and the present to generate predictions about the future'.[43] This should all sound rather familiar by now.

The third and last type of thinking I will discuss is thinking with *words*. This capacity, it seems, gives human cognition its most unique characteristic. People describe language as a tool for communication, which it is. But it is first and foremost a tool for abstraction. Some philosophers refer to this 'other' function of language as supra-communicative, but I prefer to think of it as pre-communicative; it is difficult to imagine how speech (as opposed to vocalisation) could have arisen without abstraction. Language does not merely express thoughts that we already have, it forms new ones.

An experiment by Gary Lupyan and Emily Ward reveals this function of language. They used a technique called 'continuous flash suppression' – another type of bi-stable perception, similar to binocular rivalry – in which an image presented continuously to one eye is suppressed from consciousness by a changing stream of images presented to the other eye.[44] In this experiment, subjects were shown a picture of a familiar object – such as a chair, a pumpkin or a kangaroo – in one eye, while the other eye saw a series of squiggles. The squiggles suppressed the stable image from consciousness. Immediately before looking at the squiggles and object, however, the subjects heard one of three things: either (1) the word for the suppressed object (e.g. 'pumpkin', when the object was a pumpkin), or (2) the word for a different object (e.g. 'kangaroo', when the object was a pumpkin) or (3) just static noise. When asked to indicate whether they saw something or not, the subjects were significantly more likely to report consciously seeing the stable object when the word they heard matched it than when a non-matching word or no word was paired with the image. In fact, hearing the wrong word further reduced the chances of seeing the suppressed object. The explanation of this is that 'when information associated with verbal labels matches incoming (bottom-up) activity, language provides a top-down *boost* to perception, propelling an otherwise invisible image into awareness'.[45] In other words, it increases the precision-weighting of the perceptual image.

To grasp the power of this mechanism, note that (in reality, rather than just in this experiment) we provide ourselves with such top-down labelling all the time, through the process of 'inner speech'. The boost this provides is a form of non-declarative *priming*. Intriguingly, though, priming with words has a substantially stronger effect on consciousness than priming with concrete images. A picture turns out to be worth somewhat less than a word, let alone a thousand. This is presumably because abstractions (which reside deeper in the predictive hierarchy) can achieve more than images when it comes to things like 'dogness'. Thus, for example, Lupyan found that hearing the word 'dog' has a significantly greater chance of overcoming continuous flash suppression than does merely hearing barking sounds.[46] The abstraction has greater reach. And so, when subjects are implicitly primed to attend to 'vehicles' versus 'humans' while watching a video clip, the verbal priming alters the tuning of whole neuronal populations, shifting them to be more sensitive to the presence of the one versus the other class of object.[47] Words have the power to boost whole semantic categories. Indeed, many such categories would not be thinkable – and therefore perceivable – without their verbal labels. This applies most obviously to the kind of abstract concepts we are considering in this book. Who ever saw free energy? Yet, once we can think 'free energy', we can see its workings everywhere.

If that's what one or two words can do, what happens when we start putting hundreds of them together? Take, for example, the potential of personal narratives: the abstracted stories we tell ourselves about the flow and meaning of our lives. 'Such narratives,' writes Andy Clark, 'function as high-level elements in the models that structure our own self-predictions, and thus inform our future actions and choices.'[48]

These narratives are, of course, typically *co-constructed* with others, throughout life, starting with the mother–infant dyad. This introduces the communicative function of language. The artificial manipulation of precision needn't come only from our own generative models; it can equally come from the models of others, if they have similar capacities for abstraction. Clark lists this second function of language under the heading of 'continuous reciprocal prediction', while Andreas Roepstorff and Chris Frith speak of 'script-sharing' and 'top-top control of action'.[49] In short, the abstract labelling that controls one person's precision can be communicated directly to that of another, bypassing the laborious business of bottom-up learning.

Roepstorff and Frith make this point by comparing the effects of giving verbal instructions to fellow humans to help them perform a task versus the arduous training process that is required to install sufficient knowledge in a monkey to do the same thing. In the example that Roepstorff and Frith use – a card-sorting game – humans acquire the target behaviour after a few minutes of verbal instruction, whereas it takes a full year of operant conditioning before monkeys catch on. We know from fMRI evidence that the humans and monkeys in this experiment deployed equivalent brain regions – the same cognitive set – to *perform* the actual task. It was only their method of *acquiring* it that differed.

Language entails still more. It opens the door to a whole range of precision-improving techniques, such as the observations, theories and equations featured in this book. Thanks to words (and other symbols, such as mathematical ones), models acquired over an individual lifetime can become stable objects for scrutiny and systematic improvement by others – not only our contemporaries, but over generations. With its marvellous gradations of generality and specificity, language lets us project something of the structure of the predictive hierarchy itself into consciousness. These powerful aids to cognition are not available to non-symbolic species. It is very difficult to imagine the whole of science, technology and culture without language.

My own hunch is that language evolved mainly from the PLAY drive. I explained in Chapter 5 how PLAY gives rise to the formation of social rules. Rules regulate group behaviour, and thereby protect us from the potentially excessive needs of each individual. The 60:40 rule is an innate social rule. It demands reciprocity and mutuality, and therefore facilitates the development of empathy – that is, the capacity to read other minds. It is easy to see how the forging of acquired social rules encourages complex forms of communication in order to express them, and how that in turn contributes to the emergence of symbolic thought. The pressure to develop artificial rules increased exponentially when humans shifted from the ancient hunter-gatherer lifestyles typical of all primates and began to live in permanent settlements, with planted crops and husbanded livestock. We had no evolutionary preparation for this development, which occurred a mere 12,000 years ago – none, that is, apart from the PLAY drive, which is instrumental in the formation of social hierarchies.

It is therefore of great interest to note that the cortex contributes more to PLAY than to any of the other basic emotions.[50] Its 'as if' quality is impossible to conceive without cortical mechanisms of the kind I have described in this chapter. PLAY might well be a biological precursor of thinking in general (i.e. of all virtual versus real action) and the whole of cultural life.

Perceptive readers will note that what I have said in this section also suggests something about the role of language in Freud's 'talking cure'. What is talk therapy if not a targeted intervention in one's personal narrative? In my view, psychotherapy – a form of 'continuous reciprocal prediction' – is also a form of PLAY. But now I am touching on topics that deserve books of their own.

Let me end these reflections on the cortex by considering the essential *difference* between cortical and brainstem consciousness.

Panksepp was very interested in our tendency to associate certain colours with certain feelings.[51] It is conventional, for instance, to describe red-spectrum colours as warm and blue-spectrum ones as cold. Is this arbitrary? Could the colours be reversed? Exteroceptive warmth and coldness are not, of course, visual qualities, as red and blue are. They are features of somatic sensation; but they are also closely related to what Panksepp called 'sensory affects' such as disgust, pain and surprise.[52] In several respects, the hedonic aspects of warmth and coldness embody biological value; the physical proximity of others is warm, sex is warm, fire is warm, etc. The colour spectra associated with fire and ice plausibly acquired their affective meanings in this way; as, perhaps, do the colours of ripe versus unripe fruit. Panksepp speculated that these associations might be a holdover from the evolutionary origins of perceptual qualia, a reminder of the time when our only conscious perceptions were sensory affects.

But there is a big difference between the feeling of affects and the associative valencing of externally perceived objects such as ripe fruit or hot fires. The difference is this: the affects are inborn but the valencing of external percepts is *acquired*. The visual and other perceptual qualities of percepts become 'associated' with affects in the empiricist sense.[53] We bestow values on the world. Although *some* such associations follow regularities that exist in nature – between red and fire and ripeness on the one hand and certain pleasurable feelings on the other – even those associations are acquired, and they are therefore subject to the vagaries of individual experience.

In short, such associations are *contextual*. For example, fire isn't *always* a good thing, which is presumably why warm-spectrum colours like red sometimes denote good things like sex and sometimes bad things like danger. That is also why we speak not only of 'hot' sex but also of 'hot' aggression. This principle applies in increasing degrees to the ever more individualised associations that constitute each person's subjective iconography. It follows the logic of Lisa Feldman Barrett.

Panksepp's view was that the affective values associated with some perceptual qualities are a vestige of their origins in sensory affects. But we must not lose sight of the fact that the innate connection is broken when we link perceptual qualities like redness or heat with erotic feelings, and blueness or coldness with sad ones. That is why it is readily possible to condition colour-affect associations.

Population-level regularities in these associations, which do follow nature to some degree, must surely be the basis for art forms like painting, music and dance. We appreciate their visual, auditory and somatosensory qualities – at least in part – via our sensory affects and the connotations they evoke in us. Given my definition of consciousness as felt uncertainty, it is also interesting to note the role of surprisal in aesthetic experience. The same applies to humour: a dull artwork is like a joke whose punchline is apparent in advance. But these topics are too vast to address properly here.[54]

We may conclude that perceptual qualia are different from affective qualia in that they are *not inherently* hedonically valenced; they acquire their valence in relation to affect. The light and sound waves that impinge upon the cortex all the time are not always consciously experienced;[55] but, when they are, they are felt as *context* (see p. 143–4).

What perception ultimately consists in is the computing of inferential statistics concerning the probability distributions of spike trains, and comparisons of such probabilities, all in a nested hierarchy of precision-optimising homeostats.[56] The hierarchy generates a graphic representation in our mind's eye of 'equipment-evoked responses' – *phenomena* in John Wheeler's sense. Those phenomena constitute the virtual reality displayed before (and by) each of us. The resting states of the billions of little homeostats – all embedded within each other, from the surface to the depths – represents our *confidence* in an expected context. The context is what we expect to happen beyond our Markov blanket, when we try to resolve a prioritised need. Prioritising a need triggers cognitive work rooted in an attendant belief: the expected outcome. What you perceive consciously, on the mechanism we are now familiar with, is not this expected context, but rather the nested display – over a deep hierarchy – of prioritised *deviations* from your expectations.

At the periphery of the hierarchy, such fluctuations happen all the time; it is almost impossible to predict the present in every detail. Our fluctuating confidence in the salient aspects of our sensory-motor expectations is experienced in the form of colours and tones and the like – that is, in arrangements of the exteroceptive categorical variables that our species computes. These include not only perceptions of the outside world but also flittering deviations from the expected proprioceptive states of the avatar that represents our own bodies, since our experience of our bodies is no less virtual than our experience of the outside world.[57]

Precision prediction errors in perception and proprioception (when they are salient) register as exteroceptive qualia. By contrast, changing confidence in the phenotypic belief that generated the sensory-motor process in the first place is experienced in the form of affect. No matter what project we are undertaking, we get a moment-by-moment sense of how well we are doing. You can feel it now, surely? A shifting inside you as you read these words, a flow of waxing and waning uncertainty, which might stop you in your tracks if it grows too strong. Its grounding quality, affect, is felt throughout your life and ultimately regulates everything that you do.

Our everyday experience, I suggest, ultimately consists in nothing but this.

11

The Hard Problem

The physicist Paul Davies writes:

> Among life's many baffling properties, the phenomenon of consciousness leaps out as especially striking. Its origin is arguably the hardest problem facing science today and the only one that remains almost impenetrable even after two and a half millennia of deliberation [...] Consciousness is the number-one problem of science, of existence even.[1]

I could quote many similar statements by other scientists. The 'hard problem' (as it is reverentially abbreviated) asks why and how you – 'your joys and your sorrows, your memories and your ambitions, your sense of personal identity and free will',[2] in short, your experience of existence – could possibly spring from the physiological processes that occur in brain cells. These cells are not fundamentally different from those that constitute other bodily organs. So, how do they bring 'you' into being?

The question is hardly new; it is probably the most ancient and heartfelt of all human mysteries. In past times it took the form 'How does my soul come to reside in my body?' But it was posed in its current form in 1995, by the philosopher David Chalmers. Let me quote his celebrated formulation of it:

> It is undeniable that some organisms are subjects of experience. But the question of how it is that these systems are

subjects of experience is perplexing. Why is it that when our cognitive systems engage in visual and auditory information processing, we have visual or auditory experience: the quality of deep blue, the sensation of middle C? How can we explain why there is *something it is like* to entertain a mental image, or to experience an emotion? It is widely agreed that experience arises from a physical basis, but we have no good explanation of why and how it so arises. Why should physical processing give rise to a rich inner life at all? It seems objectively unreasonable that it should, and yet it does.[3]

Chalmers's formulation owes a large debt to an earlier paper by the philosopher Thomas Nagel, 'What is it like to be a bat?' (1974). Nagel emphasised the *something-it-is-like-ness* of subjective experience. He pointed out that 'An organism has conscious mental states if and only if there is something it is like to *be* that organism – something it is like *for* the organism', and he added: 'If we acknowledge that a physical theory of mind must account for the subjective character of experience, we must admit that no presently available conception gives us a clue about how this could be done.' Nagel concluded: 'It seems unlikely that any physical theory of mind can be contemplated until more thought has been given to the general problem of subjective and objective.'

What prompted Chalmers to restate the 'general problem' in the way he did, two decades later, was the fact that brain scientists had just begun to tackle consciousness experimentally. Due to the technological advances I described in Chapter 1, they believed that questions like 'How do brain cells turn physiological processes into experiences?' could now be answered. One of the first scientists to tackle the problem experimentally was the molecular biologist Sir Francis Crick, co-discoverer of the structure of DNA. He did so in a book entitled *The Astonishing Hypothesis* (subtitled *The Scientific Search for the Soul*), which he published just one year before Chalmers's momentous declaration of disbelief. This is what Crick wrote:

> The astonishing hypothesis is that you, your joys and your sorrows, your memories and your ambitions, your sense of personal identity and free will, are *in fact* no more than the behaviour of a vast assembly of nerve cells and their associated molecules.[4]

Chalmers didn't flatly contest the claim that consciousness arises from a physical basis. As just noted, he said himself that 'it is widely agreed that experience arises from a physical basis'. He claimed only that 'we have *no good explanation* of why and how it so arises'. This repeated Nagel's assertion that 'no presently available conception gives us *a clue* about how this could be done'. Crick had asserted that a good explanation could be found: that technology now available could easily identify what he described as the 'neural correlate of consciousness'. He believed that if we isolate the anatomical parts of the brain that are necessary for consciousness, and the specific physiological functions of those parts, we could solve the mind/body problem scientifically. He recommended that we start our search by focusing on just one of the neural correlates of consciousness: the brain processes that distinguish conscious from unconscious *vision*. From there, we could presumably extrapolate to the rest of consciousness. That seemed reasonable enough. Surely there must be a neural correlate of visual experience?

As we have seen, while only *cortical* vision is conscious, the cortex can also process visual stimuli unconsciously, and so can the superior colliculi. According to Crick, therefore, the problem of visual consciousness is a simple question of asking what takes place in the visual cortex when it is processing information *consciously* that does not take place when it does so *unconsciously*, something which also does not take place in the superior colliculi. For reasons elaborated above, I believe he got off on the wrong footing, anatomically speaking; he should have focused on the brainstem rather than the cortex and on affect rather than vision. Chalmers, however, had a more fundamental reservation.

Crick's approach – which became the mainstream approach in cognitive neuroscience – elides what Chalmers calls the 'hard' part of the mind/body problem. Isolating the neural correlates of consciousness is the 'easy' part. It merely identifies the specific brain processes that *correlate* with experience; it does not explain how they *cause* it. That is the hard part of the problem: how and why do neurophysiological activities *produce* the experience of consciousness?[5] In other words, how does matter *become* mind? According to Chalmers, we neuroscientists may be able to explain how neural information is processed in the brain while we are having visual experiences, but that does not explain how those brain processes *turn into* experiences. As John Searle, another major philosopher of mind put it: 'How does the brain get over the hump from electrochemistry to feeling?'[6] This question is equally perplexing when framed the other way round: how do immaterial things like thoughts and feelings (e.g. deciding to make a cup of tea) turn into physical actions like making a cup of tea?[7]

The extent of this explanatory gap,[8] as philosophers call it, is well illustrated by the 'knowledge argument', which goes something like this.[9] Imagine a congenitally blind neuroscientist named Mary who knows everything there is to know about the neural correlates of vision. Although she can explain all the physical facts of visual information processing – right down to the cellular level, including the impact of light waves on the photosensitive rods and cones, and how these waves are converted into nerve impulses, and how those impulses are propagated via the lateral geniculate body to the cortex, and how they are further processed there by neatly arranged columns of neurons, organised in vast numbers into a variety of information-processing modules spread far and wide throughout the cortical mantle, the multiple specialised visual processing streams of which are well understood – still she would not know *what it is like* to experience vision. Being blind since birth, she would know nothing about the experienced qualities of redness and blueness – for example – which are, after all, the actual stuff of conscious seeing. This is not only because she herself has never experienced such qualities, but also because nothing in her anatomical and physiological *knowledge* about the neural correlates of vision explains what it is like to see. If she were to suddenly acquire the gift of sight, she would learn something utterly new about vision – something that none of her mechanistic understanding prepared her for. *The physical facts therefore do not explain why and how there is something it is like to see.* They only explain why and how the brain decodes visual information: how *it* sees, not how *you* see. This supposed irreducibility to the physical basis of what philosophers call 'qualia' – the something-it-is-like-ness of subjective experience – is the hard problem. That, according to Chalmers, is 'the central mystery of consciousness'.[10]

This perceived irreducibility has led great minds through the ages, not least of them the physician-philosopher John Locke, to the conclusion that conscious experiences *are not part of the physical universe*.[11] Since the qualia of experience never-theless clearly do exist, these thinkers relegated them to some *non-physical* dimension of reality, which they (or many of them) described as 'epiphenomenal'.[12] It was Locke who pointed out in his 'inverted spectra argument' that it is logically possible for someone to experience the quality of blueness as I experience the quality of redness, and still call it 'red' just like me, even though what they consciously see is what I would call 'blue'. The upshot of his argument is that *it would make no difference* if they did so – these two extremes of the colour spectra could readily be interchanged in relation to their experienced qualia. Therefore, according to Locke, consciousness is not explained by its physical correlates; the relationship between the physical mechanism (in this case, the relative wavelengths of light) and the psychological qualia is not a causal one.

Consciousness thus becomes something that just follows alongside the chain of physical events in the brain – a sort of by-product – with no impact upon the causal structure. Locke was writing in the seventeenth century, but this is not an archaic view. It is widespread, and not only among philosophers. As two respected cognitive scientists recently put it: 'Personal awareness is analogous to the rainbow which accompanies physical processes in the atmosphere but exerts no influence over them.'[13]

If conscious experiences play no role in the workings of the physical world, why do they exist? What does awareness add to the information processing that happens anyway, unconsciously? What is the point of becoming aware of brain processes if your awareness has no influence over them?[14]

This is not true for vision alone. One of the most compelling bits of experimental evidence for the view that consciousness is a mere epiphenomenon of brain processes is the observation by Benjamin Libet to the effect that one's subjective decision to initiate a movement is preceded (by 300 milliseconds or so) by measurable brainwaves that announce the onset of the movement you believe you initiated.[15] In other words, the physical initiation of the movement (in the brain) begins before you consciously decide to move; it is not really 'you' who initiates it. This finding is widely taken to prove that conscious choice – 'free will' – is an illusion. If free will does not exist, then what remains of consciousness? As we saw, Crick's answer was that you, your joys and your sorrows, etc., do not 'in fact' exist. If only we had an adequate physical account of such qualia, says Crick, then we could *explain them away*.

I have always found it impossible to accept either the argument that conscious qualia *exist in some parallel universe* or the argument that they *do not exist at all*. And so should you, because you are your consciousness. To claim that you are like a rainbow that exerts no influence over your physical body is patently absurd. It is equally absurd to claim that 'you' are in fact nothing more than the behaviour of neurons and therefore do not really exist. These claims are contradicted by every moment of your experience.

The claim that 'you' do not exist also flies in the face of one of the most famous conclusions of all philosophical thought. After a lifetime of contemplation as to what he could be *absolutely certain of*, René Descartes (in his 'philosophy of doubt') famously arrived at the conclusion that the *only* thing we need have no doubt about is the fact that we exist: 'I think, therefore I am.' In other words: you experience, therefore you exist.

This is Chalmers's hard problem: since an experiencing self clearly does exist, how do we accommodate it within our physical conception of the universe?

I am not a philosopher. As you have seen, my interest in the neurology of consciousness arose and developed independently of the philosophical literature, and I confess to finding much of what I have read of it rather baffling. Nevertheless, in this book I have tried to provide a natural scientific response to the 'hard problem'. I do not claim to have dispelled every mystery in the metaphysics of consciousness. As we'll see, deep questions remain. But I do believe that I have shown how natural processes, unfolding according to their various necessities, can (over evolutionary time) generate something very like our private worlds of experience.

How does this bear on the problem as Chalmers articulates it? How does the account I have provided bridge the explanatory gap between the inner and outer worlds?

Let's start from a basic observation: the laws governing mental functions such as perception, memory and language are *abstracted* from both subjective and objective data. To take one example, Ribot's Law, a scientific abstraction about long-term memory, accounts for the observable fact that your *internally* experienced memory of what happened ten years ago is more securely consolidated than what happened ten minutes ago. This is why elderly people are more likely to forget recent events than remote ones. The same applies to *externally* observed, physiological memory traces; the ten-year-old ones are more securely consolidated than the ten-minute-old ones (to exactly the same degree).[16] So, the temporal gradient in Ribot's Law is neither psychological nor physiological; it is both. Likewise, consider Miller's Law, a scientific abstraction about short-term memory. It accounts for the experienced limitation of your capacity to hold things in mind: at any given moment, you can only retain seven units of information (plus or minus two). The duration of short-term memories can also be measured: they typically last between fifteen and thirty seconds.[17] The very same capacity limitation can be observed physiologically: it involves neurotransmitter depletion.[18] Miller's Law, therefore, like Ribot's, is both psychological and physiological.

Laws like these, which can in principle be quantified and therefore expressed mathematically, explain the dual manifestations of an abstracted function called 'memory'. Surely the same applies to 'consciousness'. If the phenomena of consciousness are natural things (what else could they be?), they, too, must be reducible to laws.

This is not a radical or idiosyncratic conclusion. The whole of cognitive science, which has dominated my field since the late twentieth century, is based upon it. Recall that in Chapter 1, I wrote:

> The most interesting thing [about cognitive 'information processing'] is that it can be implemented with vastly different types of physical equipment. This casts new light on the physical nature of the mind. It suggests that the mind (construed as information processing) is a *function* rather than a structure. On this view, the 'software' functions of the mind are implemented by the 'hardware' structures of the brain, but the same functions can be implemented equally well by other substrates, such as computers. Thus, both brains

and computers perform *memory* functions (they encode and store information), and *perceptual* functions (they classify patterns of incoming information by comparing them with stored information) as well as *executive* functions (they execute decisions about what to do in response to such information).

However, I immediately went on to say:

This is the power of what came to be called the 'functionalist' approach, but it is also its weakness. If the same functions can be performed by computers, which presumably are not sentient beings, then are we really justified in reducing the mind to mere information processing?

This brings us to the heart of the hard problem. According to Chalmers, *the hard problem of consciousness cannot be solved by 'functionalist' explanations*:

The easy problems are easy precisely because they concern the explanation of cognitive abilities and functions. To explain a cognitive function, we need only specify a mechanism that can perform the function. The methods of cognitive science are well-suited for this sort of explanation, and so are well-suited to the easy problems of consciousness. By contrast, the hard problem is hard precisely because it is not a problem about the performance of functions. The problem persists even when the performance of all the relevant functions is explained [...] What makes the hard problem hard and almost unique is that it goes *beyond* problems about the performance of functions. To see this, note that even when we have explained the performance of all the cognitive and behavioural functions in the vicinity of experience [...] there may still remain a further unanswered question: *Why is the*

performance of these functions accompanied by experience?
A simple explanation of the functions leaves this question
open [...] Why doesn't all this information processing go on
'in the dark', free of any inner feel?[19]

In Chapter 4, I outlined the failed attempts to specify a mechanism that can perform the function of consciousness. Chalmers's response to the first of them conveys the tenor of his attitude to them all:

Crick and Koch suggest that [synchronised gamma] oscillations are the neural *correlates* of experience. This claim is
arguable [...] but even if it is accepted, the *explanatory* question remains: Why do the oscillations give rise to experience?
The only basis for an explanatory connection is the role they
play in binding and storage, but the question of why binding
and storage should themselves be accompanied by experience is never addressed. If we do not know why binding and
storage should give rise to experience, telling a story about
the oscillations cannot help us. Conversely, if we *knew* why
binding and storage gave rise to experience, the neurophysiological details would be just the icing on the cake. Crick and
Koch's theory gains its purchase by *assuming* a connection
between binding and experience, and so can do nothing to
explain that link.[20]

Here is Chalmers's critique of the second major functionalist account of consciousness, the 'global workspace' theory of
Newman and Baars (1993):

One might suppose that according to Baars, the contents of
experience are precisely the contents of the workspace. But
even if this is so, nothing internal to the theory *explains* why
the information within the global workspace is experienced.

The best the theory can do is to say that the information is experienced because it is *globally accessible*. But now the question arises in a different form: why should global accessibility give rise to conscious experience? As always, this bridging question is unanswered.[21]

Chalmers concludes:

The usual explanatory methods of cognitive science and neuroscience do not suffice. These methods have been developed precisely to explain the performance of cognitive functions, and they do a good job of it. But as these methods stand, they are *only* equipped to explain the performance of functions. When it comes to the hard problem, the standard approach has nothing to say [...] To explain experience, we need a new approach.[22]

I must be clear about what Chalmers is saying here. He is saying that the methods of cognitive science can only explain the performance of cognitive functions, and they do a good job of it. By 'cognitive functions' he means things like perception, memory and language. So, he is saying that *consciousness is not a cognitive function*. Why is he saying that? The answer is: because he is not talking about an abstracted *function* called 'consciousness' but rather about the *experience* of consciousness – about *what it is like* to perceive or remember.

The hard problem would be trivial if all it boiled down to was the fact that your own individual experience is not the same as human experience in general.[23] If that were the hard problem, all we would need to do in order to solve it would be to take the single-witness experiences of lots of individuals, average them, find the common denominator, and explain *that* in functional terms. Psychologists do this sort of thing all the time. That is what I did when I researched confabulation: I started with the subjective experiences of one individual, Mr S, generalised them by studying the equivalent experiences of other patients like him, and then abstracted the common denominator. This approach revealed a functional principle about confabulation, namely that it makes the patients feel better; confabulation serves a 'wishful' function.

But Chalmers is not merely saying that subjective phenomena are connected with a single point of view. He is asking: why should *any* function be accompanied by experience? Accordingly, he writes:

> Why is it that when electromagnetic waveforms impinge on a retina and are discriminated and categorised by a visual system, this discrimination and categorisation [function] is experienced as a sensation of vivid red? We know that conscious experience *does* arise when these functions are performed, but the very fact that it arises is the central mystery. There is an *explanatory gap* (a term due to Levine, 1983) between the functions and experience, and we need an explanatory bridge to cross it. A mere account of the functions stays on one side of the gap, so the materials for the bridge must be found elsewhere.[24]

It's worth spelling out what this 'explanatory bridge' must achieve, so that we do not shift the goalposts. In an article published shortly after Friston and I submitted our (2018) article, three Czech colleagues announced that a solution to the 'hard problem' was imminent, and they predicted it would come from the Free Energy Principle:

> There is no clear understanding or general consensus among philosophers and neuroscientists about the function of consciousness. This is one of the main reasons why consciousness still represents such an elusive problem, which has its roots in the fact that there has never been an articulated function of

consciousness based on and supported by a unifying brain theory. Such a unifying theory is emerging in the form of the predictive coding framework ... based on the ideas of Hermann von Helmholtz that the brain is mainly a predictive inference machine.[25]

However, these authors added a warning: 'Unfortunately, the hard problem of consciousness will probably never completely disappear because it will always have its most committed supporters'. They continued:

We believe that there is little hope that the most committed proponents of the hard problem would be completely satisfied with [any] conclusions of empirical science, since the core argument of the hard problem is aimed at the endeavours of empirical science in the first place.

Sure enough, when Friston and I received our first peer reviews from the *Journal of Consciousness Studies*, that is exactly what one of them said about our article: 'The hard problem (following Chalmers) is a metaphysical problem and as such it is not open to being "solved".' (Friston wrote to me afterwards: 'I get the impression that the hard problem is not there to be solved; it is there to be revered.')[26]

To be fair to Chalmers, he is not responsible for the philosophical prejudices that troubled our Czech colleagues. Consider the closing sentence of his 1995 article: 'The hard problem is a hard problem, but there is no reason to believe that it will remain permanently unsolved.'

So, let's be clear about what the explanatory bridge he asks for must achieve. I will quote Chalmers's own criteria. In his demolition of the first functionalist theory, he wrote: 'The question of why [gamma synchronised] binding and storage should themselves be accompanied by experience is never addressed.' So, when assessing whether the explanatory gap has been bridged by any functionalist theory of consciousness, you must ask yourself: *Has the question as to why XYZ function should itself be accompanied by experience been addressed?* In relation to the second theory that Chalmers rejected, he added: 'Why should global accessibility give rise to conscious experience? As always, this bridging question is unanswered.' So, when assessing any other theory, you must ask yourself: *Has the question as to why XYZ function should give rise to conscious experience been answered?* And that, apparently, is all.

But if that's all, then why does Chalmers add that 'a mere account of the functions [of experience] stays on one side of the gap, so the materials for the bridge must be found elsewhere'? Why must an account of the functions fail to bridge the gap? Re-reading what he says in this connection, as quoted above, it becomes apparent that he is conflating *two different kinds of explanatory gap*. So, we must be clear which gap we are speaking about, if we are going to bridge it.

The first one is an explanatory gap between signals propagated from the retina and sensations of vivid red – that is, between physiological events and psychological events. I must point out that *both these types of event are in fact experienceable*. You can experience physiological events such as observing signals propagated from the retina just as easily as you can experience psychological events such as sensing vivid red. Neither of these events can be explained – and still less explained away – by the other. *They are two ways of observing the same thing.* When I (introspectively) experience myself as existing, is the mental thing that exists a different thing from the bodily Mark Solms I see in the mirror? And is the Mark Solms in the mirror a different thing from the anatomical Mark Solms that is seen in an MRI scanner? I do not understand why so many people (even philosophers) talk about the brain as though it were somehow exempted from reality *as we experience it.*[27] This can only be because when they say 'the brain' they mean something else. They don't mean the brain that we see and touch; they mean something *abstracted* from our experience of it – something in the nature of a *functional* system.

The first explanatory gap is therefore one located between *two different kinds of experience* associated with two different observational perspectives.[28] This is analogous to *hearing* thunder with your ears and *seeing* lightning with your eyes. People do not remark: 'It is widely agreed that thunder arises from lightning, but we have no good explanation of why and how it so arises.' This is because they do not believe that lightning *produces* thunder in the way that livers produce bile. They accept that they are two manifestations of the same underlying thing.[29] This applies equally to the different ways of experiencing visual information processing: from the outside or the inside. From the outside (if you are a scientist with the right equipment) you see signals propagated from the retina;[30] from the inside you see vivid red.

The second explanatory gap is located between experiences (of both kinds) and their underlying causes. It is the gap between things you can experience, such as vivid reds and optogenetic scans of activated neurons, and things you cannot experience, such as quantum fields in themselves. It is, in short, a gap between the first-person and third-person perspectives. To take a third-person perspective on my own experience is to abstract myself from the experience, and to experience it no longer. This perspective concerns neither the brain as it looks nor the mind as it feels, but rather the forces *explaining* why and how it looks and feels as it does. That is the perspective I have taken in this book.

I am not sure it is obvious to everyone that Chalmers, in the quotation above, is referring to both of these gaps simultaneously. This can give rise to confusion. He takes the gap between two types of experience (exterospective observations versus introspective ones) and conflates it with the gap between experience in general and its underlying functional mechanisms. This makes for a considerably wider chasm to cross.

It seems to me that most people assume that what Chalmers calls 'function' is synonymous with what he calls 'the physical', and that 'experience' is therefore only and always something non-physical. But if 'the physical' means the observable body and its organs, including the brain, then it is experienced no less than the psychological is.* Moreover, psychological experiences, when abstracted, reveal the functional mechanism of such experiences, giving rise to psychological laws.

The same applies to physiological laws: they too are abstracted from experience – from the observable physiological data. These two types of laws, being made of the same abstracted (explana-tory) stuff, namely functions, are not as difficult to reduce to each other as the two categorically different types of experienc-ing are. Thus, Miller's and Ribot's Laws are both psychological and physiological, and that is why they can be reduced to uni-fying equations. Failing to make explicit the intermediate steps

---

* I remember well my experience of the first brain operation I attended. One side of the patient's head was shaved and a yellowish-brown liquid was applied to it. Then a sweeping curve was drawn onto the scalp with a marker pen. The skin was cut through by a scalpel run deftly along that line, then flapped back over the head, exposing the skull. Tiny rivulets of blood trickled over the domed surface. These were calmly mopped. Then, four large holes were drilled into the bone, by hand. I feared the drill bit would slip into the brain. Next, a slender saw blade was pulled through the holes, and dragged from side to side. I smelled burning. This procedure was repeated four times, and the released door of bone was then removed and placed in a dish. Next came the dura mater, a thick membrane within the doorway, which bled profusely as it was cut through. Each vessel was patiently cauterised. The procedure reached its climax. The dura mater was peeled away, like a veil, and there it was: the pale pink convolutions of the cortex itself, with some snake-like vessels nestling in its grooves. A feeling of *awe* overcame me. It was as though I were entering a cathedral. Now came the task which I myself would one day have to perform: a brief series of focused conversations with the patient, as an electrode probed the glistening jelly-like convolutions of his brain, before the ensuing incision was made – an incision into the mind …

between experiential data and explanatory mechanisms exaggerates the hard problem and makes it appear harder than it is. You don't have to try to imagine how one kind of experience could produce another kind of experience: you just need to come up with a mechanistic theory that explains both sets of phenomena, of whatever perspectival modality. Then you can test the predictions it gives rise to.

Chalmers concedes that the hard problem can be solved. In fact, he sets out three principles upon which its solution should be based. Before I introduce you to these principles (one of which we must save for Chapter 12), I need to make clear that Chalmers believes we cannot solve the hard problem, ever, if we try to do so by *reducing* 'experience' to what he calls 'physical processes'. He writes:

> We are already in a position to understand certain key facts about the relationship between physical processes and experience, and about the regularities that connect them. *Once reductive explanation is set aside*, we can lay those facts on the table so that they can play their proper role as the initial pieces in a *non-reductive* theory of consciousness, and as constraints on the basic laws that constitute an ultimate theory.[31]

To me 'non-reductive' explanation is a good thing if it means we can forgo the impossible task of reducing psychological phenomena to physiological ones, or vice-versa. Psychological phenomena cannot be reduced to physiological ones any more than lightning can be reduced to thunder. Lightning does not *cause* thunder; the two phenomena *correlate* with each other. This is the easy problem. Therefore, we must reduce both phenomena to their respective mechanisms, so that we can reduce these mechanisms to a common denominator, without violating the laws of physics. This is the hard problem.

Yet Chalmers seems to mean something different by the term 'non-reductive' explanation. For him, it means we cannot reduce experienced psychological phenomena to functional laws, period.

He continues:

> A non-reductive theory of consciousness will consist in a number of *psychophysical principles*, principles connecting the properties of physical processes to the properties of experience. We can think of these principles as encapsulating the way in which experience arises from the physical. Ultimately, these principles should tell us what sort of physical systems will have associated experiences, and for the systems that do, they should tell us what sort of physical properties are relevant to the emergence of experience, and just what sort of experience we should expect any given physical system to yield. This is a tall order.

That is his first principle. He calls it the 'Principle of Structural Coherence'. It entails structural coherence between 'the properties of experience' and 'the properties of physical processes'. This seems to refer to two classes of *phenomena* which, in my view – like his – cannot be directly reduced to each other. But Chalmers also speaks of 'sort[s] of physical systems' and their 'associated experiences'. This seems to refer to the relationship between *functional* systems in general and experiences in general (whereby the latter 'emerge from' the former, or the former 'yield' the latter). That is why, for Chalmers, the Principle of Structural Coherence encapsulates the way in which 'experience arises from the physical'.

Again, he conflates physical phenomena and 'physical' (i.e. functional) causes. Unless we unpack the intermediate steps, we are bound to conclude that experience does *not* arise from the physical. The same conflated meaning of 'the physical' can be seen in the typical functionalist theories I mentioned before.

Those theories looked for the key to psychological experience in X Y or Z *physiological* mechanism: the synchrony of gamma oscillations; the binding and storage of integrated sensory and frontoparietal activity; the activation of the cortex by the intra-laminar nuclei of the thalamus; thalamocortical 're-entrant loops'; etc. Using the analogy of thunder and lightning, this conflation requires us to explain auditory phenomena by way of the functional mechanisms of vision. It is only in this sense that Chalmers is justified in saying that 'a mere account of the func-tions stays on one side of the gap, so the materials for the bridge must be found elsewhere'.

To explain psychology in relation to physiology, we must abstract ourselves from the observed phenomena *of both kinds* (i.e. we must infer functional mechanisms, of both kinds) – and then abstract ourselves from the two sets of abstractions to see the unifying common denominator. In doing so, we must situate ourselves at an *equal distance* from them both (i.e. we must infer sufficiently deep mechanisms to account for the functions of both psychology and physiology); only then can we reconcile the phe-nomena and their underlying mechanisms with each other.

· I don't want to be too formulaic about this. Real science pro-ceeds in less orderly ways. We start with an insight or discovery somewhere within the cascade of causality, and then we fill in the gaps. But by skipping over the intermediate steps, we arrive at the requirement of 'non-reductive' solutions to the hard problem which are impossible to obtain. Remember what Chalmers's ques-tions were in the first place: Why should physical processing *give rise to* a rich inner life?; how and why do neurophysiological activ-ities *produce* the experience of consciousness? These questions cannot be answered non-reductively. That is because of the way they are framed.

Understandably, therefore, Chalmers concludes that a reduc-tive solution, per se, is impossible:

I suggest that a theory of consciousness should take experience as fundamental. We know that a theory of consciousness requires the addition of *something* fundamental to our ontology, as everything in physical theory is compatible with the absence of consciousness [...] A non-reductive theory of experience will specify basic principles telling us how experience depends on physical features of the world. These *psychophysical* principles will not interfere with physical laws, as it seems that physical laws already form a closed system. Rather, they will be a supplement to a physical theory. A physical theory gives a theory of physical processes, and a psychophysical theory tells us how those processes give rise to experience. We know that experience depends on physical processes, but we also know that this dependence cannot be derived from physical laws alone. The new basic principles postulated by a non-reductive theory give us the extra ingredient that we need to build an explanatory bridge [...] This position qualifies as a variety of dualism, as it postulates basic properties over and above the properties invoked by physics. But it is an innocent version of dualism.

And so, in the end, Chalmers says something spooky. Because he believes that 'experience' cannot be reduced to 'the physical' (as he uses the word, which means both 'physiological' and 'functional'), he is compelled to conclude that *experience is not part of the known physical universe*. This would appear to be pure dualism, of the not-so-innocent kind bequeathed to us by Descartes and Locke.[32] That is why Chalmers claims that consciousness requires the addition of 'something fundamental to our ontology'; something *non-physical* 'over and above the properties invoked by physics'; something to 'supplement' physical laws; etc. I hope it is clear why I disagree with him on this score. In light of all I have told you, I cannot agree that 'everything in physical theory is compatible with the absence of consciousness'.

But there is more. Having taken his stand against reducing 'experience' to 'the physical', Chalmers goes on to say (as if to illustrate the conflation) that they are nevertheless dual aspects of something else. The 'something else', according to him, is *information* – which, I agree, is not something physical in the *physiological* sense but which, in my view, is something physical in the *functional* sense of statistical mechanics. This move introduces Chalmers's second principle. He calls it the 'Double-Aspect Principle', which he says is more basic than the Principle of Structural Coherence:

> The basic principle that I suggest centrally involves the notion of *information*. I understand information in more or less the sense of Shannon (1948). Where there is information, there are *information states* embedded in an *information space* [...] An information space is an abstract object, but following Shannon we can see information as *physically embodied* when there is a space of distinct physical states, the differences between which can be transmitted down some causal pathway [...] The Double-Aspect Principle stems from the observation that there is a direct isomorphism between certain physically embodied information spaces and certain *phenomenal* (or experiential) information spaces. From the same sort of observations that went into the Principle of Structural Coherence, we can note that the differences between phenomenal states have a structure that corresponds directly to the differences embedded in physical processes; in particular, to those differences that make a difference down certain causal pathways implicated in global availability and control. That is, we can find the *same* abstract information space embedded in physical processing and in conscious experience. This leads to a natural hypothesis: that information (or at least some information) has two basic aspects, a physical aspect and a phenomenal aspect. This has the status of a basic principle that might underlie and explain the

emergence of experience from the physical. Experience arises by virtue of its status as one aspect of information, when the other aspect is found embodied in physical processing.

Note that Chalmers here twice describes information as something 'abstract'. Once the two meanings of 'the physical' are distinguished, therefore, it becomes clear that information has the same ontological status in relation to its physiological and psychological aspects as electricity has in relation to lightning and thunder. Lightning and thunder are dual aspects of electrical discharge just as signals propagated from the retina and sensations of vivid red are dual aspects of information processing; they are different equipment-evoked responses, different phenomenal manifestations of a unitary causal process. And this causal process is *physical* in the explanatory sense.

Chalmers even cites the physicist John Wheeler in this connection. As I explained before, Wheeler introduced a 'participatory' interpretation of quantum mechanics, in terms of which the observed universe manifests itself in response to the questions that are asked of it. That is why the very same thing can take complementary forms, such as occurs with waves and particles. Wheeler's interpretation is that the phenomenal form that the experienced universe takes depends upon the way in which it is observed or measured – that is, upon one's perspective on it. A memory, for example, can be experienced either as a reminiscence or as an activated neuronal trace, depending upon the equipment you use to observe it. The form it assumes is in the eye of the beholder.

But Chalmers sees it differently. For him the complementarity inheres *in the thing observed*, not in the observer or the act of observation:

The laws of physics can be cast in terms of information, postulating different states that give rise to different effects without actually saying what those states *are*. It is only their position in an information space that counts. If so, then information is a natural candidate to also play a role in a fundamental theory of consciousness. We are led to a conception of the world on which information is truly funda-mental, *and on which it has two basic aspects, corresponding to the physical and the phenomenal features of the world.*[33]

So, Chalmers thinks the dual aspects of 'the phenomenal fea-tures of the world' and 'the physical' are *in* information itself – at its source – not in the equipment of the participant observer.[34] This is like thinking that sweetness is intrinsic to the molecular structure of glucose.[35] Perhaps this distinction does not matter where the thing observed is the observer, such as when you see your own body in a mirror.[36] And perhaps it doesn't matter so much in the end that Chalmers conflates two meanings of 'the physical' (viz., the phenomenal-physiological and the mechanical-functional levels), because he nevertheless reduces both of them to 'information'.

Two things do still present a problem, however. Firstly, for Chalmers, the properties of experience and those of the physical are not *reducible* to information. They are both *inherent in* information. Secondly, for him, they are both inherent in *all* information. That is why he states elsewhere in his article that 'phenomenal properties are the internal aspect of information', which he contrasts with physical (supposedly non-phenomenal) properties, which are its 'external aspect'. This is the strange pass to which his dualism leads him.

He concludes:

This could answer a concern about the causal relevance of experience – a natural worry, given a picture on which the physical domain is causally closed, and on which experience is supplementary to the physical. The informational view allows us to understand how experience might have a subtle kind of causal relevance in virtue of its status as the intrinsic nature of the physical.

I hope this book has convinced you that the causal relevance of experience for complex self-organising systems like ourselves is anything but subtle.

All the same, we still need to explain why and how experience arises lawfully from physical mechanisms. Chalmers asks: 'Why is the performance of these [physical] functions accompanied by experience?' Like Nagel before him, he defines experience in terms of 'something-it-is-like-ness'. To Chalmers, this is the essence of subjectivity: 'the internal aspect of information', which is, in turn, 'the intrinsic nature of the physical'. He goes on to argue that subjectivity cannot be reduced to something non-subjective. That is why, for him, subjective experience must be inherent *in* information.

But this would seem to imply that the experienced quality of being something (of being information) is a fundamental property of all things. Why does Chalmers think that *experienced* being is 'the internal aspect of information' in general? If we argue, as I have done here, that subjectivity is merely an observational perspective, where a subject is simply the being of a certain kind of object, then we are free to say that for many objects – in fact, for most of them – there is *not* 'something it is like' to be them. *The vast majority of things do not possess something-it-is-like-ness when they are considered from their own subjective point of view.*

Is there something it is like to be a single cell or a plant? What about a rock? Is there something it is like to be a computer or the internet? What about a thermostat? Chalmers says: 'It is undeniable that some organisms are subjects of experience.' Does that mean that *only* organisms (or only *some* organisms) are subjects of experience? He concedes that these are open questions, and that it may be the case that experience only arises at a certain level of complexity or with a certain type of information processing:[37]

> An obvious question is whether *all* information has a phenomenal aspect. One possibility is that we need a further constraint on the fundamental theory, indicating just what *sort* of information has a phenomenal aspect. The other possibility is that there is no such constraint. If not, then experience is much more widespread than we might have believed, as information is everywhere. This is counter-intuitive at first, but on reflection I think the position gains a certain plausibility and elegance. Where there is simple information processing, there is simple experience, and where there is complex information processing, there is complex experience. A mouse has a simpler information-processing structure than a human, and has correspondingly simpler experience; perhaps a thermostat, a maximally simple information processing structure, might have maximally simple experience? Indeed, if experience is truly a fundamental property, it would be surprising for it to arise only every now and then; most fundamental properties are more evenly spread. In any case, this is very much an open question.

If you accept Chalmers's view that experience is a fundamental property of everything, then there is nothing to explain; and if experience is everywhere and eternal then there may be little to fear, either. For the price of just one wild speculation, we brush aside the hard problem and award ourselves immortality, shaking off the fear that our existence might *depend* on anything as if it were only a bad dream.

Would that it were so. Unfortunately, as we saw in earlier chapters, experience arises only under special conditions. It sits in a precise location among the streams of information that course through our pitifully fragile beings. It performs particular tasks and then it disappears. To all appearances, it is no more basic to the structure of reality than any other variety of equipment-evoked response. I am very sorry. If you do not accept Chalmers's speculation, then we have to answer this question: why is there something it is like to be some things but not others? If not all objects are subjects, and not all subjects are sentient, then how does it happen that sentient subjects *sometimes* come about?

I find Chalmers's expansive proposal about information intriguing, but I think it is far more plausible that '*we need a further constraint* on the fundamental theory, indicating just what *sort* of information has a phenomenal aspect'. Defining this constraint has been the main task of this book. We have seen that everything pivots on the aim and purpose of the information processing. When it comes to consciousness, this means *minimising entropy*. But it is more complicated than that: one also needs a Markov blanket. Then it entails minimising *your own* entropy. Moreover, it entails doing so across myriad categorical parameters in unpredicted contexts. The physical facts I have outlined in this book reveal that consciousness does not inhere in all information, but rather in a certain type of information processing: a complex form of the self-evidencing type.

If only some things, or only organisms, or only some organisms, are subjects of experience, then consciousness cannot be a fundamental property of the universe. There certainly was a dawn of life: there is an abundance of empirical evidence for this. And life emerged a long, long while after the Big Bang. Therefore, there surely was a good deal of time before consciousness existed. This assumption alone – that there was a dawn of consciousness – obliges us to find a physical explanation for it. If there was a dawn, there must have been something prior to consciousness that explains it. The alternative notion – that consciousness preceded life and the universe – doesn't fit the facts as they appear to be, and moreover it sounds unhelpfully like the idea of God. Unless you invoke ideas like that, consciousness must have arisen from, and therefore must be part of, a non-conscious physical universe.

This applies even within ourselves. We saw in previous chapters that perception and cognition are not necessarily accompanied by consciousness. In fact, the scientific evidence suggests that perception and cognition are *mostly* unconscious. I cited Kihlstrom's and Bargh and Chartrand's classic review articles in this connection. After reviewing the evidence, these scientists came to the conclusion that we are unconscious of our psychological acts 'most of the time'.

There was much in the empirical findings that these authors reviewed which deserved our attention (especially the manner in which unconscious intentionality derives from conscious learning from experience),[38] but I want to remind you of just the upshot: events happen and exist inside you whether you are conscious of them or not, including *psychological* events. We could have arrived at this – in a sense obvious – conclusion from many different starting points, but that doesn't matter. If most 'moment-to-moment psychological life' carries on without experience, then why can't thermostats and the internet be unconscious? If the being of the non-declarative basal ganglia of the brain is not 'intrinsically experiential' (to use Chalmers's term), then how can all information be experiential?

Clearly, subjectivity in general must be made of something non-experiential. This is not a metaphysical point about things-in-themselves versus things-as-we-experience-them; it is an empirical fact. Moment-to-moment psychological life *as I experience it* is not the totality, by any means, of my psychological life. This puts experience into perspective. As Freud taught us, not everything in psychological life is conscious. And we can only fathom the nature of consciousness by discerning the particular job it does. That is, its function.

I keep asking: what does awareness *add* to information processing

– to the information processing that happens anyway, unconsciously? What is the point of becoming aware of physical processes if your awareness has no influence upon those processes? This question, it seems to me, is the same as Chalmers's core question: 'Why is the performance of these functions accompanied by experience? [...] Why doesn't all this information-processing go on "in the dark", free of any inner feel?'

In my view, the question only arose because Chalmers, following Crick, sought the function of consciousness in the wrong place. The fundamental form of consciousness is not something cognitive, like vision; rather, it is something affective. In that sense, and that sense alone, Chalmers was right to imply that consciousness is not a *cognitive* function: the primary function of consciousness is not perceiving or remembering or comprehending but *feeling*.

How can the function of feeling go on 'in the dark', without any feeling? We can legitimately ask why vision is accompanied by experience. Vision does not require consciousness, and neither does any other cognitive process. But feeling does.[39] Some theorists, it is true, claim that drives and affects and emotions are not *necessarily* conscious. This is because different people mean different things by those words. That is why I have used the word 'feeling': if there are such things as unconscious emotions, 'feeling' denotes their counterparts that are conscious. There can be no such thing as a feeling that you do not feel. 'Unconscious feeling' is an oxymoron.

That is why I have focused the scientific arguments in this book upon feeling. *In order to solve the hard problem of consciousness, science needs to discern the laws governing the mental function of 'feeling'.* This is not just a matter of words. I marshalled considerable evidence to show that feeling is the foundational form of consciousness, its prerequisite. I also explained both physiologically and mechanistically the difference between felt and unfelt needs and showed that feelings have concrete consequences. This enabled me to conclude in Chapter 9 that 'consciousness is not merely a subjective perspective upon the "real" dynamics of self-organising systems; it is a function with definite causal powers of its own'.

The perplexity in Chalmers's questions – 'Why is the performance of these functions accompanied by experience?' and 'Why doesn't all this information processing go on "in the dark"?' – disappears when we realise that 'experience' is not intrinsic to vision, and to information processing in general. It is intrinsic only to the specific form of information processing that generates feeling.

That, in my view, is why there is *not* something it is like to be a thermostat or the internet. That is the crucial pertinence of the questions 'Why is there something it is like to be some things but not others?' and 'Why (and how) does it happen that sentient subjects *sometimes* come about?'

These questions must be answered in mechanistic terms. What is the function of feeling? Can feeling perform its function without experience? If the blind visual neuroscientist Mary were an *affective* neuroscientist, would her knowledge of everything there is to know about the function of feeling not explain why (indeed predict that) it feels like something?[40] The same applies to Locke's inverted spectra. Is it logically possible for someone to experience as excruciatingly painful everything that I experience as exquisitely delicious, and for this to make no difference? Certainly not. This is because feelings really *do* something – and they greatly increase our chances of survival in the process.[41]

It is easy to recognise the explanatory gap between visual information processing and visual awareness. But the same kind of gap does not exist between the function of feeling and the experience of it. Some philosophers will claim there still is such a gap. They will point out that 'affective zombies' are *conceivable*.[42] I suggest that this will be based in the historical preoccupations of their discipline rather than reasonable perplexity, and I appeal to them to consider the issue afresh.

Concerns about an 'explanatory gap' would never have arisen if we had begun our quest by asking why and how feelings arise, rather than by looking for a neural correlate of consciousness in the visual cortex. The biological function of feelings like hunger is nothing mysterious; and their something-it-is-like-ness is not especially difficult to explain. Just follow the logic of free-energy minimisation where it leads for self-organising systems like us. Given our multiple needs, complex and perilous environments, wide choice of possible actions and ability to perform only one or two of them at any given time, we should *expect* to have an inner world, built for the purposes of deliberation and choice. And what should we expect to fill it? What else but a dynamic range of valuative qualities, centrally including confidence weightings, which tag and measure our various incommensurable needs as they arise, along with the salient features of the environment in which they must be met.

Consider the following statement by Chalmers. Would he have said such things in the first place if he had been talking about affective – rather than cognitive – functions?

> This is not to say that experience *has* no function. Perhaps it will turn out to play an important cognitive role. But for any role it might play, there will be more to the explanation of experience than a simple explanation of the function. Perhaps it will even turn out that in the course of explaining a function, we will be led to the key insight that allows an explanation of experience. If this happens, though, the discovery will be an *extra* explanatory reward. There is no cognitive function such that we can say in advance that explanation of that function will *automatically* explain experience.[43]

I can say in advance that explanation of the function of feeling will automatically explain experience. I do not see how an adequate natural scientific account of it can fail to do so. Starting with thermodynamics, we arrive – surprisingly easily – at a qualified and agentic subjectivity, one whose most urgent priorities are weighed for a moment, felt, and then transformed with (one hopes) due circumspection into ongoing action. That, I submit, is what it is like to be anything that there is something it is like to be.

Does this clear up every last mystery about the nature of consciousness? Here I must admit to a residue of discomfort. How strange it would be if all that has ever been experienced – if the very knowability of the universe itself – depended on the mechanisms I have described. My sensibility rebels at the thought. Then again, it is strange that we exist, or indeed that anything does. One can't escape the fact of contingency. That can be unsettling even in more workaday contexts – the close call, the realisation that, but for some unsuspected circumstance, we might not have made it. And perhaps it *should* be unsettling: viewed from a certain perspective, it is precisely the brush with death that our feelings evolved to prevent. I am reminded again of Mr S, whose subjectivity worked overtime to conceal the precariousness of his existence following his unsuccessful surgery. He is not the only one. Presented with an account of our being that leaves it as tenuous as mine has, perhaps an impulse towards wishful denial is only natural.

There's more to it than that, though. All explanations must take something as given, and therefore inexplicable within the theory. Every story must end somewhere. For me, the trail ends with information, which is undoubtedly puzzling stuff, and with self-organisation, which is positively uncanny. In the account of consciousness that I have given, everything springs from a system's drive to exist. Our minds are woven from order itself, which emerges spontaneously from chaos as in Friston's experiment, and then defends itself against the onslaughts of entropy. How can this be the basis for our existence? What *is* order, that it has such powers to conjure us out of the inanimate darkness before and beyond us? These questions exceed the scope of this book. For all I know, they elude enquiry altogether. Nevertheless, I would love to know the answers to them.

Short of that further revelation, I hope what I have provided is still of value. It is, after all, something that many have doubted could ever be possible: an account of sentience in terms of other things that we know exist in the physical world. There are still many hard problems, but perhaps not *the* hard problem – anyway, not in quite the form we have grown accustomed to. And if sometimes even I doubt that consciousness can be what I have said it is – if I do not feel 100 per cent certain of my account – I take comfort in the fact that niggling uncertainty, with no prospect of final illumination, is exactly what my theory would predict.

12

Making a Mind

As a child, it seemed to me there was no point in doing anything. No matter what I did, I was going to disappear forever. My consciousness was inextricable from my brain, and judging by all the evidence, that is a strictly time-limited thing. This caused me great distress. The only way I could find out of the nihilistic hole was to try to understand what consciousness *is*. If I did so sincerely and with concerted effort, then at least I would not have wasted my brief allocation of existence. I would have spent it on the only problem worth solving in the circumstances. This way of proceeding also held out the hope – remote but not impossible – that by understanding what consciousness is, I might elude its confines. I might somehow find a way to escape the solipsistic bubble of existence, to find a way of contextualising 'being' within some bigger picture. Through this, I confess, I hoped I might even find some alternative to the terrifying logical implication of mortality.

And so, I took the path that culminated in this book. Along the way – during my neuropsychological education and psychoanalytical training – I took comfort in the discovery that logic itself was the product of a limited instrument. I gained direct knowledge of how much 'moment to moment psychological life' occurred outside of conscious awareness and voluntary control. This put thinking in its place, as it were. I also saw how some neurological patients were unaware of the most obvious truths, due to loss of specific parts of their brains. If people like Mr S are driven to such incorrect assumptions by specific limitations of their mental instruments, perhaps the same applies to me. What if we are all, like them, lacking the machinery that would, if we had it, enable us to come to radically better conclusions about ourselves and our place in the universe? If we were all endowed with only a sense of hearing, we might think that reality consisted in something as ethereal as sound waves, and we would have no conception of the visible and tangible world of mental solids. Since we would all be constrained by the same incomplete evidence, we would all reach the same wrong conclusions. By the same reasoning, if we had an additional bit of brain – say, five lobes per hemisphere instead of the usual four – perhaps we would know something about the nature of things that we are currently missing.

That may be so. But then it is equally true that the unknown facts could turn out to be *more* depressing than the ones we are currently labouring under. Some might argue that the more brain power one has, the more one's place in the universe turns out to be worse – not better – than one thought. I am not sure that that viewpoint is justified. Either way, we have no choice but to do the best we can with what we have. The rules of science demand that we test our conjectures against the best evidence we can find, and be ready to reject the hypotheses that it disconfirms. The rules of life demand the same of everyday experience. As things stand, there is scant evidence to support the hypothesis that my sentient being will outlive my mortal flesh. We appear wholly dependent on our alarmingly fragile brains.

On the basis of this premise, we may suppose that consciousness did not exist on earth before brains evolved – and perhaps only when vertebrate brains evolved; therefore, about 525 million years ago. I suspect it arose in rudimentary form before that; that a precursor of affect gradually became felt affect, with no sharp dividing line between them, in tandem with the evolution of increasingly complex organisms with multiple competing needs. What emerged with the evolution of cortex was *cognitive* consciousness – that is, the additional capacity to contextualise affect exteroceptively and hold it in mind.[1] In any event, sentient being cannot have existed before nervous systems existed. The inner, subjective form of consciousness cannot have existed in the absence of its outer body. Therefore, on the basis of the evidence I marshalled in Chapter 6, we may conclude that consciousness as we know it requires the existence of something which looks like the PAG, or its immediate evolutionary precursor, together with its adjacent equipment in the midbrain decision triangle and reticular activating system.

As you read these words, perhaps some glimmer of doubt is arising in you. Certainly, it flickered into my awareness sometime between February and July 2018. I had begun meeting regularly with a small group of physicists and computer scientists who shared, or were at least sympathetic to, my view that recasting consciousness as feeling might open the way to a physical account of it.[2] As we proceeded with our deliberations, I found myself taking a view that might not surprise you now that you have read this book – but it did initially surprise me. The unsettling thought that occurred was this: consciousness as we know it from the inside does not necessarily entail the existence of something that *looks* like the PAG. It only requires the existence of something that *functions* like it.

David Chalmers, unlike many of his followers, believes that the hard problem can be solved. In fact, in the very article in which he first outlined the problem, he sketched a possible solution to it under the heading: 'Outline of a theory of consciousness'. His theory is not very well known, probably because few scientists accept it. It is based on three principles, two of which I introduced in the previous chapter. They were the Principle of Structural Coherence and the Double-Aspect Principle. I took issue with both, but I was also sympathetic to them. In a way, this book simply nuances them. I told you that I would introduce the third principle now, in the final chapter. This is the Principle of Organisational Invariance.

This principle simply states that any two systems with the same fine-grained functional organisation will have qualitatively identical experiences. If the causal patterns of neural organisation were duplicated in silicon, for example, with a silicon chip for every neuron and the same patterns of interaction, then the same experiences would arise. According to this principle, what matters for the emergence of experience is not the specific physical makeup of a system, but *the functional pattern of causal interaction between its components*. This principle is controversial, of course. John Searle, among others, has argued that consciousness is tied to a specific biology, so that a silicon isomorph of a human would not be conscious.[3] I believe that the principle can be given significant support by the analysis of thought experiments, however.

Chalmers outlines one such thought experiment.[4] It revolves around the notion of two information-processing systems with identical functional organisations. In one of them, the organisation is produced by a configuration of neurons (as it is in the natural brain) and in the other it is produced by a configuration of silicon chips (as it might be in an artificial brain). As we know, natural human brains are capable of experience. The question Chalmers raises is whether the same applies to an exact functional replica of the human brain. Here comes the imaginary experiment:

> The two systems have the same organization, so we can imagine gradually transforming one into the other, perhaps replacing neurons one at a time by silicon chips with the same local function. We thus gain a spectrum of intermediate cases, each with the same organization, but with slightly different physical makeup.[5]

The question that arises here is whether replacing a natural neuron with an artificial one will make any difference to what the natural brain experiences, and vice-versa. The argument that ensues is that the neuronal brain will continue to have the same experiences as it goes through all the tiny intermediate steps, up to and including the point at which it becomes a complete silicon brain replica. (The same applies the other way round, of course.) This shows – says Chalmers – that silicon brains, too, are capable of experience:

> Given the extremely plausible assumption that changes in experience correspond to changes in processing, we are led to the conclusion [... since no changes in processing occurred during the neuron-to-silicon transition] that *any two functionally isomorphic systems must have the same sort of experiences.*[6]

Chalmers grants that 'some may worry that a silicon isomorph of a neural system might be impossible for technical reasons. That question is open. The invariance principle says only that *if* an isomorph is possible, then it will have the same sort of conscious experience.' He concludes:

> This thought experiment draws on familiar facts about the coherence between consciousness and cognitive processing to yield a strong conclusion about the relation between physical structure and experience. If the argument goes through, we know that the only physical properties directly relevant to the emergence of experience are *organizational* properties.[7]

I cannot find any flaw in this logic. It corresponds roughly to my own argument to the effect that a single functional organisation (e.g. Mark Solms) may assume two different appearances, one introspectively and the other extrospectively. My disagreement with Chalmers arises only from what he does with this argument when he develops the theory that *all* information has a subjective aspect – and is therefore conscious. I said (quoting Chalmers) that 'we need a further constraint on the fundamental theory, indicating just what *sort* of information has a phenomenal aspect'. That is all we really disagree about. In this book I have identified the special sort of information processing that I think does have a phenomenal aspect, and I have explained why and how this occurs. Whether Chalmers accepts this 'constrained' version of his Double-Aspect and Structural Coherence principles or not,[8] he and I still agree that any two functionally isomorphic systems must have the same sort of experiences. In other words, if one of two functionally isomorphic systems has phenomenal experiences then *the other must have them too.*

Before examining the implications of this conclusion, let us consider the scientific worry that a silicon isomorph of a neuronal system might be impossible for technical reasons. Chalmers says 'that is an open question'. These words were written in 1995. The question is no longer open. In a series of studies published between 2012 and 2016, several groups of researchers showed that it is possible to create an artificial interface between the brain and the spinal cord which enables paralysed animals to move their affected limbs by replacing spinal neurotransmission with radio signals.[9] Marco Capogrosso's group, for example, reported the following:

> Rhesus monkeys (*Macaca mulatta*) were implanted with an intracortical microelectrode array in the leg area of the motor cortex and with a spinal cord stimulation system composed of a spatially selective epidural implant and a pulse generator with real-time triggering capabilities. We designed and implemented wireless control systems that linked online neural decoding of extension and flexion motor states with stimulation protocols promoting these movements. These systems allowed the monkeys to behave freely without any restrictions or constraining tethered electronics.[10]

These monkeys had (unilateral) damage to upper motor neurons in the corticospinal tract at the level of the thoracic spine, causing paralysis of one leg. 'As early as six days post-injury and without prior training of the monkeys, the brain–spine interface restored weight-bearing locomotion of the paralysed leg on a treadmill and overground.'[11] The crucial point in this study is that the 'brain–spine interface' (i.e. the radio signals that restored communication between the cortex and the lumbar spine, bypassing the corticospinal lesion) is nothing other than *an artificial isomorph* of the kind that Searle worried may be 'impossible for technical reasons'. The radio signals replaced the neuronal signals in exactly the manner that Chalmers envisaged his imaginary silicon chips might replace neurons. In Capogrosso's study, the radio signals performed the same function that the missing corticospinal neurons usually perform.

It is true that I am talking about corticospinal neurons that perform only *motor* functions (rather than cognitive ones), but it is important to note that the neurons in question are 'pyramidal' neurons that originate in layer 5 of the neocortex. They are therefore morphologically identical with the type of neurons that are widely described in the predictive-coding literature as processing both 'prediction' and 'error' signals in both perceptual and active inference. Neurons of this type are found throughout the cortex, including in the hippocampus and prefrontal lobes. The fundamental point, therefore, is this: the function of a pyramidal neuron can be performed by an artificial isomorph (in this case, its function as an interface between neocortex and muscle can be performed by radio waves). And to ram the point home, a fully functioning silicon isomorph of a pyramidal neuron has recently been engineered.[12]

Now I want to focus your attention on the crucial aspect of the study I have just described. What the intracortical array implanted over the motor cortex of the rhesus monkeys recorded was *information*. The information produced by the cortical activity was a *message* that the monkey brains tried to send to the monkey legs, using corticospinal pyramidal neurons. What the array of microelectrodes did was encode this message into radio waves and then transmit the very same information via an artificial medium to the lumbar spine, which then transferred it back to its natural medium, to produce the intended movement. In other words, the function that the radio waves performed artificially was an information-processing function that is normally performed naturally by corticospinal pyramidal neurons. The same principle applies to the recently engineered silicon neurons, although at the time of writing they have not yet been used in such experiments.

To make this point more broadly, let me explain that the same thing can be done with other information-processing functions of the brain, including uniquely human cognitive functions. I am not only saying that this can be done, it *has* been done. In a landmark study published in 2012, Brian Pasley and colleagues recorded electrical activity from intracortical arrays that were placed over the auditory cortex of fifteen human volunteers who were undergoing brain surgery for medical reasons.[13] During the operations, the researchers transmitted words through loudspeakers or headphones which the patients consciously listened to. The team studied the resultant EEG recordings and crafted an algorithm – a computational model – to map the sounds the patients heard onto the EEG patterns they had recorded. The model matched each speech sound with its corresponding pattern of brain activity. The researchers then reverse-engineered the process: starting with the patterns of brain activity (i.e. the information), they showed that it was possible to use the computer model to reconstruct the words that the brain had heard.

Here the computer algorithm plays an equivalent role to the one that the radio waves played in the previous experiment, but a more complex one. It models *artificially* the very same information that is modelled naturally by the brain, then it generates the corresponding sound that is yielded by that model.

What these scientists did with words can be and has been done with visual imagery too. Shinji Nishimoto and Jack Gallant showed that it is possible to develop computer algorithms which decode – from fMRI recordings alone – what the visual cortex is seeing, and then turn those patterns of brain activity back into moving pictures.[14] In this way, starting with the fMRI recordings alone, it is possible to generate reasonable approximations of the visual images that caused the brain activity.

The same thing has even been done with dreams. The study that did this is nothing short of breathtaking:

> Visual imagery during sleep has long been a topic of persist-ent speculation, but its private nature has hampered objective analysis. Here we present a neural decoding approach in which machine-learning models predict the contents of visual imagery during the sleep-onset period, given measured brain activity, by discovering links between human functional mag-netic resonance imaging patterns and verbal reports with the assistance of lexical and image databases. Decoding models trained on stimulus-induced brain activity in visual cortical areas showed accurate classification, detection, and identi-fication of contents. Our findings demonstrate that specific visual experience during sleep is represented by brain activity patterns shared by stimulus perception, providing a means to uncover subjective contents of dreaming using objective neural measurement.[15]

It is a small step – a very small step – to go from this to decod-ing a person's thoughts, operationalised as inner speech or mental imagery. Christian Herff and colleagues – who showed that 'con-tinuously spoken speech can be decoded into the expressed words [as written text] from intracranial electrocorticographic (ECoG) recordings' – say almost casually that this will be the next step in their research programme: 'The Brain-to-Text system described in this paper represents an important step toward human-machine communication based on imagined speech.'[16]

These are far-reaching developments. To take just one apparent implication: if, in the manner envisaged by Chalmers's thought experiment, it is possible to encode the contents of cortical processes[17] into an artificial medium and then decode them back again into physiological processes that produced them, then it should in principle be possible to decant any expanse of cortical processing into an artificial device, and file it away for future use. This makes it possible, in principle, for the contents of an individual's cortex to outlast their mortal flesh. And if this is possible for, let us say, a mouse, then it should also be possible for a human being – since the basic architecture is the same.

When I first became aware of this possibility, I was not overly excited by it.[18] Even if it were possible to store artificially the entire contents of a person's long-term memory, still we would not have copied that person's mind. We would have achieved little more by doing this than we can already achieve when we keep someone's body alive artificially, for example in a persistent vegetative state or coma. If I did this for you, I would not be keeping *you* alive, I would only be keeping *it* alive. On this analogy, if (or perhaps I should say when) it becomes possible to store artificially the contents of declarative long-term memory – the totality of any expanse of cerebral cortex's individualised functional parameters – then we would still only be storing an 'it' rather than an 'I'. My reasoning took the view, on scientific grounds, that cortical memory parameters are not intrinsically conscious. There would be *nobody at home* in those long-term memory systems – no subjective presence to palpate the traces with feeling. That is why I dismissed the whole artificial intelligence enterprise: 'Unless it is possible to design a computer that has *feelings* […] it will probably never be possible to design a computer with a mind […] The problem of the mind is therefore probably not a problem of *intelligence*.'[19]

Perhaps you can see where I am going with this.

By 2002, when I wrote those words, I knew that consciousness was fundamentally affective. (Looking over my publications, I appear to have come to this understanding gradually since 1996.[20]) Back then, however, I took the view that affect had nothing to do with information processing. I equated *cognition* with information processing and I thought of affect as something more intrinsically biological. Now, looking back – although these distinctions still make sense to me – I no longer understand why I considered neuromodulation to be more 'biological' than neurotransmission. Moreover, I no longer see why I assumed that affect is not a form of information processing.

Perhaps I just didn't know enough about information science. Specifically, I didn't understand that *entropy* in statistical mechanics is equivalent to *average information* (see Chapter 7) – which means that the patterns of brain activity that the above-mentioned researchers recorded from the cortex are ultimately patterns of *homeostatic deviations* from neuronal resting states. In other words, I didn't understand that cortical activity, too, is entropic – that it shares this common currency with affect – and that cognitive work (i.e. predictive work) is *anti*-entropic, and therefore part of the same economy. I only began to understand this in 2017, while preparing my Closing Remarks at that fateful London congress.

Now, strange as it seems even to myself – an avowed non-cognitivist, a 'biologist of the mind' – I am obliged to take a different view of artificial intelligence. Today, I would go so far as to say that, unless it is possible to make a conscious machine, we will not have solved the hard problem.[21] If the special form of information processing I have proposed here really is the causal mechanism of consciousness, then it *must* be possible to produce artificially a conscious mind with it. That outcome is the only scientifically tenable way to prove the concept. Even if the technology by which a hypothesis can be tested doesn't yet exist (as was the case in Freud's day), still the hypothesis must be formulated in a way that renders it testable in principle.

And when it comes to the hypotheses I have outlined in this book, the requisite technology *does* exist.

As a neuropsychologist, the biological mechanism of consciousness that I formulated in Chapter 6 is the best I can come up with. It makes plausible sense of all the available neurological and psychological data. Much else that is useful could be added by studying more closely the physiological means by which the PAG processes the multiple converging inputs it receives, and how exactly decisions regarding 'what to do next' are made when it interfaces with the superior colliculi. But such research would still belong under the heading of what Chalmers calls the 'easy' problem.

In Chapter 9, with the considerable help of Karl Friston, I reduced the psychophysiological functions in question to formal statistical mechanics. The resultant equations are broad-brushstroke cartoons of the underlying *causal mechanisms* of consciousness, in both its psychological and physiological manifestations. The reverse-engineering of these abstractions will tell us whether they can solve the hard problem. As things stand currently, there are too many gaps between the broad brushstrokes I have delineated and the practical implementation of the mechanisms they describe; and these missing steps will only come into view fully as we try to instantiate them.

I am looking forward to undertaking this work with the enlarged team of physicists, computer scientists, biomedical engineers and neuroscientists that I have assembled. By the time this book appears in print there will hopefully be progress to report. For now, all I can tell you is how I *envisage* we will proceed. Everything that follows, therefore, is subject to the evolving input and advice of my specialist colleagues as we advance with the next stages of our project.

Our starting point, it seems to me, will be quite different from that of most traditional 'AI' research. Firstly, what we are trying to engineer is not intelligence but rather *consciousness*.[22] I hope it is clear by now that I do not conceive of consciousness as being particularly intelligent – at least not in its elementary form. And it seems like a good strategy to try to engineer consciousness in its most elementary form, not only for convenience but also for ethical reasons that I will discuss shortly. Secondly, we will not be trying to engineer a device that does anything practical – chess-playing or voice recognition or the like – the achievement of which functional criterion will signal that we have reached our goal. Rather, we will be engineering a self-evidencing system with no *objective* end in view other than remaining the means to that end. In other words, we will be trying to make a being that has no aim and purpose other than to carry on being.[23]

So, we will *start* with something similar to what Friston created: an unconscious self-organising system equipped with a Markov blanket (and therefore with sensory, active and internal states) which automatically models the world on the basis of sensory samples, minimising the effects of entropy upon its functional integrity by improving its generative model. That is, in line with Friston's Law, it will measure its own expected free energy and act accordingly. This will turn it into a prediction machine. It will now maintain an increasingly complex, hierarchical generative model of its organisation in relation to prevailing external states, even though, like the leak-plugging Eve Periaqueduct, it will have no explicit task other than to plug its system's leaks.

This will be a lifelike system, but it will not be alive. Although consciousness evolved in living organisms, the purpose of this experiment is to show that it can also be produced *artificially*, by reverse-engineering its functional organisation. There is every reason to believe that we can create such a (self-organising) system, because it has been done before. Although this first-stage system will not be conscious, it will already have subjective values: its valuable states (from its own point of view) will be the states that minimise its free energy, and it will do whatever is necessary to maintain those states for as long as possible. The proto-intentionality this implies will be relatively easy to verify, as it can readily be inferred from the observable behaviour of the system.

To move to the second stage, at which I believe the *precursors* of affect will first appear, we need to make our system more complex. Specifically, we need to give it multiple needs. This can (and will initially) be done through computer simulations, but a more realistic approach requires us to eventually *embody* the system physically, as a robot, and make its capacity to proactively minimise Friston free energy contingent upon an external source of Gibbs free energy.[24] If we did that, the system would have to *represent* the Gibbs free energy supply in terms of Friston free energy, and then bind it through effective information-processing work. This would make the maintenance of its external energy supply a core responsibility of the self-organising system. (Of course, the information-processing work done by every computer relies upon electricity, but this is normally provided gratis by an external agent.) The hardware of the computer would thus become the system's body, even before we place it in a robot, and the values of the self-organising system would oblige it to act – within its model of the world – in ways that best enable its body to absorb the requisite external energy.

Initially in simulation, we could design various motor contraptions that physically enable the system to tap into the external energy source – or better still, sources (plural) – in order to recharge its internal battery. We could also shift the activated sources to different locations in the computer's environment (using a reasonably complex regimen that the system would need to learn). This, in turn, would require us to equip the system with a perceptual apparatus. In a similar vein, we could burden it with a thermoregulatory mechanism, to stop it overheating if it works too hard. Since this would threaten the physical integrity of the body upon which the self-organising system depends, it would have to represent and regulate this homeostat too. Here we could also incorporate a fatigue parameter, in terms of which the system's information-processing efficiency becomes compromised (gradually, progressively) when it is active for too long, without rest, and this parameter could interact in complex ways with the other demands we place upon it. We could add a pain parameter, triggered by damage (e.g. wear and tear) to specific aspects of the computer's physical integrity, such as the joints and surfaces of its wheels and arms. This makes it easy to envisage an anxiety parameter, tied not only to the perceived danger of incurring pain but also to *impending* existential risk relating to the other parameters. To create more realistic emotions, it will be necessary also to have the system compete with other agents (similar to itself) for its energy resources, and to face threats from them, and to have attachment needs of them, and to form alliances with them. And so on. This will instantiate the important emotional problem of *conflicting* needs.

Many draft versions of the system would fail to master the onerous demands envisaged above, and would accordingly expire. It would therefore be necessary to monitor what works best and to preset the successful predictive codes and their associated precisions into each new generation of the system phenotype. (This is artificial 'natural selection', necessitated by the fact that the envisaged artificial systems will not physically self-generate or reproduce.)[25]

At last we come to the third stage. Now the multiple needs of the system must be flexibly prioritised through precision optimisation on a *contextual* basis. For example, the quantum of externally available energy could vary in relation to thermoregulatory and fatigue and anxiety thresholds, so that optimal precision weightings during one circadian or quasi-seasonal epoch would differ from those in another. Likewise, the system could be allowed to adapt to a complex configuration of environmental risks and opportunities and then the parameters could be changed, thus creating novel (unpredicted) environments, and a need for longer-term forward planning.

No doubt, this will again result in the expiration of many iterations of our self-organising system. For the same reasons as before, therefore, what works best will have to be artificially propagated through successive generations of the phenotype. We need not be overly concerned about this, because the systems that do not survive will presumably lack the essential ingredient we are attempting to engineer here – namely the capacity for sentience.

Consciousness (in the sense of uncertainty *felt by the system*) ought to arise only at this third stage of the experiment. Self-organising systems imbued with self-preservative values (as they are by definition) already contain the raw ingredients for the function we call affect; that is, they already register subjective 'goodness' and 'badness' of a kind that only applies *to* and *for* them, and can therefore only be felt *by* them. On my hypothesis, this inherently subjective property is what we call hedonic *valence*. Stated in formal terms, increasing free energy is an existential crisis for any self-organising system. Such systems must therefore develop internal models of themselves in relation to their worlds, which confer on them the capacity for self-preservative action. This qualifies as proto-intentional behaviour. With increasing complexity, the system increasingly has to prioritise its intentions flexibly and on a contextual basis, and then hold one such intention 'in mind' (in a short-term memory buffer) to guide its unfolding choices in uncertain environments. It will also need to plan ahead and 'remember the future' over longer timescales.

At this point, only those systems that can compartmental-ise the different error values which contribute to the free energy calculations, and can flexibly modulate their attendant precision weightings, are likely to survive. These values must be treated as categorical variables; that is, they must be differentiated from each other on a *qualitative* rather than a quantitative basis, and one of these qualities must be prioritised at every given moment, and then implemented and evaluated in terms of the system's fluctuating levels of *confidence*. Then, I predict, the inherently subjective, inherently valenced, inherently existential, inherently qualitative, inherently intentional internal states of the system will become what we call 'feeling'.

I can see no reason to come to any other conclusion than that a sufficiently complex, well-tuned, self-evidencing system which is capable of finding itself in this state could *feel* it. The feelings in question would not be *the same* as human feelings, or mammalian feelings, or animal feelings in general, but they could be feelings nonetheless. Now, the question becomes: how will we know if or when the artificial self feels itself to be in this internal state? How can we prove it?

If you want a rough-and-ready sense of the current mainstream view about something, the obvious place to look is Wikipedia. Here is what it has to say about artificial consciousness:

> Qualia, or phenomenological consciousness, is an inherently first-person phenomenon. Although various systems may display various signs of behaviour correlated with functional consciousness, there is no conceivable way in which third-person tests can have access to first-person phenomenological features. Because of that, and because there is no empirical definition of consciousness, a test of presence of consciousness in artificial consciousness may be impossible.[26]

This is a big challenge. In fact, it is the problem of other minds all over again. As I said before, we are not trying to engineer an AI device that does something practical, the achievement of which observable criterion will signal that we have reached our goal. The system we must engineer has no objective goal other than to survive, and, in particular, to survive in unpredicted environments. But if it achieves this objective criterion, how will we know whether it has used *feeling* to do so? That is what I expect – I predict that it will use feelings – but feelings are inherently subjective things, so, as Wikipedia blithely reminds us, 'there is no conceivable way' in which we can demonstrate their presence objectively.

Once again, the rules of science come to our rescue. Science does not require absolute proof. This latitude does not apply only to the mechanism of consciousness; it applies to everything. All that the rules of science demand are that one puts up a reasonable hypothesis (which takes account of all the available evidence) and then tests it in a way that provides a realistic chance for it to be falsified. So, in the present case, how will we know when the hypothesis that our self-evidencing system is conscious has been falsified? As we discussed in the previous chapter, it is conceivable that 'affective zombies' might exist who *behave* as if they are conscious but are not conscious. How, then, can we ever know whether they *really are* conscious?

The traditional answer to this question is the Turing test, which is a very simple test that requires you to interact with the system via a computer monitor (i.e. without *seeing* it) and to decide on the basis of its verbal statements alone whether it has a mind or not. If you think you are interacting with a real mind (technically, if 30 per cent of judges after five minutes of interaction think they are), then the machine has passed the test.

Here is an example that I published with Oliver Turnbull in 2002.[27] He interacted – via the internet – with someone or something called ELIZA who was purporting to be a psychotherapist. His task was to decide whether she was a real person or not:

ELIZA: How do you do? Please tell me your problem.
Oliver: Why can't I be happy?
ELIZA: Do you think you should be able to be happy?
Oliver: I think I deserve more happiness than I have at the moment.

ELIZA: Do you really think so?
Oliver: Yes, life doesn't need to be so cruel.
ELIZA: You seem quite positive.
Oliver: I suppose I am, at least some of the time.
ELIZA: Is that why you came to me?
Oliver: I came because I need help.
ELIZA: Is that the real reason?
Oliver: Actually, I came because I heard you were a computer and not a real therapist.
ELIZA: Do computers worry you?
Oliver: Can I ask a question?
ELIZA: Whether you can ask a question depends more on you than on me.
Oliver: The question is: how do mind and brain relate?
ELIZA: I'm not sure I understand you fully.
Oliver: It's the hard problem – how does mental life arise from physiological events?
ELIZA: Please go on …

I hope you guessed that ELIZA is not a real person, she is a chatbot – a computer program – and a very simple one at that. Nevertheless, a surprisingly large number of people who interact with her believe she really is a psychotherapist, and many of them find her to be helpful.[28] Clearly, there is something wrong with the Turing test. As Oliver Turnbull and I pointed out, the main thing wrong with it is that it is a *behaviourist* test of intelligence; it is not actually a test of *consciousness*. It is true that the intelligent behaviour of a machine can sometimes be indistinguishable from that of a person, and when it is, the machine should be treated as equal (or even superior) to us in terms of intelligence.[29] That is, people and computers do indeed both display 'intelligence'. But the problem of philosophical zombies raises another matter; it is precisely *not* a matter of behaviour or intelligence but rather of the difficult fact that, when it comes to consciousness, appearances can be misleading.

Several other formal tests have been proposed over the years – some of them specifically aimed at testing for consciousness – but they are all just as unsuitable for our purposes as the Turing test, mainly because they assume you are testing for cognitive rather than affective consciousness.[30] We are trying to engineer something much simpler than that: a mind whose cognitive processing is no more complex than is required for it to feel its own dwindling energy supplies or too-hot-ness.

The good thing about the Turing test is that it circumvents prejudice.[31] There is a danger that we humans will assume a priori that something that looks like a 'mere machine' cannot possibly be conscious. This could become a self-fulfilling prophecy. There is a long history of humans displaying such bigotry, and it continues to this day. Here I am not particularly referring to prejudice on grounds of race, gender or sexual orientation, which are bad enough, but to the assumption that children born without a cortex must be unconscious, as must non-human animals. If many people – even respected neuroscientists – are inclined to believe that rats, our fellow mammals, equipped with essentially the same midbrain anatomy that we are, and with a cortex, whose every act is congruent with the hypothesis that they are conscious, nevertheless lack it, then what hope is there that they will accept that our artificial self is sentient, no matter how much evidence we provide?

We will have to see what happens. For my own part, all I can do is lay out my predictions explicitly and say how I intend to test them. My main prediction is that the second-stage system described above will not be able to survive in novel environments but that the third-stage system (or some versions of it) will. This is my operational criterion for *voluntary* activity (see p. 100 for a definition of 'voluntary'). Moreover, I predict that the two different outcomes will coincide with some critical aspect of the functioning of a *need-prioritisation* (i.e. precision optimisation) mechanism, which only the third-stage system will possess. What exactly the critical feature will be within this broad mechanism can be determined only on a trial-and-error basis. In short, we must identify the artificial 'neural correlate of consciousness' of our system – its affect selection mechanism and the mechanism by which it holds a prioritised affect in mind and uses it to qualify the uncertainty in an unfolding action sequence. The identification of such a feature will enable us to manipulate it, in much the same way as we have done with the components of the vertebrate brain that we concluded are responsible for *our own* consciousness (which remains the one and only form of consciousness that we can verify directly, empirically, i.e. in our own cases, due to the problem of other minds).

For example, we may confidently predict that *damaging* the neural correlate of consciousness in our system will obliterate its consciousness in much the same way that lesioning the parabrachial complex does with us vertebrates, or rather that the damage will obliterate its voluntary behaviour in the same way that lesions to the vertebrate PAG do with us. Likewise, we may predict that *stimulation* (i.e. enhancement) of this critical component of the system will facilitate volitional activity. We can also, of course, readily expect that internal activity recorded from it will predict not only external events but the attendant goal-directed voluntary behaviours, and that different aspects of this recorded activity will correspond with different aspects of the observable behaviours.

What I most hope, once we have identified the neural correlate of consciousness in our artificial system, is that this component will prove to be sufficiently differentiable from other components of its functional architecture – specifically those that are responsible for implementing the adaptive behaviours that feelings normally give rise to – that we can manipulate the putative feelings *uncoupled* from their adaptive consequences. Remember what I said in Chapter 5 about the distinction between subjective psychological motives and objective biological design principles.

For example, sexual behaviour is typically motivated by the pleasure it produces rather than the reproductive imperatives that attached biological 'reward' to procreative acts over evolutionary time. What I have in mind here is something analogous to what one observes in addicts, who are motivated to perform work in order to achieve desired feelings, even though the feelings themselves bestow no adaptive advantage upon the system in terms of its underlying design principles.[32] Something similar is observable in animals (zebrafish, for example) that exhibit conditioned place-preference behaviour for locations where they received opiates, cocaine, amphetamines and nicotine – that is, *hedonic rewards* which bestow little if any adaptive advantages and can actually cause harm.[33] If the equivalent can be demonstrated for our system, it seems to me, this will be weighty evidence for the presence of subjective feeling – evidence which could then be cross-validated by way of the causal manipulations (such as artificial lesioning and stimulation) and recording techniques envisaged above.

Naturally, all of this will still be subject to the problem of other minds. But the same applies to you and me. I can never know for sure whether you are conscious. In the end it comes down to converging findings and the weight of evidence. There may never be consensus on this score, just as there is no consensus today as to whether hydranencephalic children and non-human animals are subjects of experience. People who do not accept that these fellow beings are conscious probably never will accept that an artificial 'being' feels anything, no matter how much supporting evidence is produced. For the rest of us, there should always be room for doubt. Speaking for myself: if I find that I am substantially *undecided* as to whether our artificial self is conscious or not, that would be a remarkable outcome.

Should it be done? When our team first began to contemplate a research project of the kind I have just described, we were quickly beset by ethical concerns. What is the purpose of making such a machine? Who might benefit from it, and in what ways, and at what cost, to whom? In short, what are the risks?

Many current AI projects and applications are commercially motivated. Should we accept research funding from someone seeking to profit financially from our project? On what basis might an artificially conscious device become profitable to someone?[34] It is understandable that commercially motivated persons (in the legal sense, which includes corporations) might wish to replace human labour with artificial units of production where the latter might either be more efficient than us – including more intellectually capable – or more 'willing' to perform relentless and monotonous tasks than us. Even this motive is ethically questionable insofar as it leads to concerns about diminished employment prospects for human beings, but at least there can be no question of *exploitation* of non-conscious machines.

This cannot be said in the case of our project. To the extent that artificial consciousness might be used for financial gain, to that extent we are at risk of facilitating a new form of slavery. This would be a gross failure of empathy, just as it always was. I therefore cannot imagine any ethical justification for developing sentient robots through a commercially funded research programme – or even for developing them at all for such expedient purposes, if this raises the prospect that they might be so exploited.

Our concern for the welfare of putatively conscious machines should, of course, extend beyond the fear that they might be exploited for pecuniary reasons. As soon as machines acquire consciousness (even the most rudimentary forms of raw feeling) then more general questions concerning their potential for *suffering* necessarily arise. The dominant tradition in Western ethical theory, 'consequentialism' (i.e. the notion that the consequences of one's conduct are the ultimate basis for judgements about its rightness or wrongness), is deeply concerned with pain and suffering. By creating artificial feeling beings, we therefore enter into the jurisdiction of that kind of ethical calculus.

The question of *rights* is more controversial. Once machines become sentient beings, do the issues that are currently debated under headings like 'human rights', 'animal rights', 'children's rights', the 'right to life' and so on, become applicable to them too? Should conscious machines have rights to 'life' and liberty? In fact, the concept of 'robot rights' is already established, and the issue has been considered by the Institute for the Future in the US and by the Department of Trade and Industry in the UK.[35]

By way of example, I will mention just two issues that apply to conscious (as opposed to intelligent) robots. As we try to create them – even to learn whether it is possible to do so – are we justified in deliberately placing them in putatively distressing situations, in order to demonstrate aversive responses? (This question often arises in experiments with animals.) And assuming we can create sentient machines, on what grounds would it be ethical to switch them off again? These two examples could easily be multiplied.

Alongside such ethical and moral questions, there are practical questions to consider, some of them very important, even existentially important. For example, since computers are already more intelligent than us in some limited respects, is it not possible that those which are both highly intelligent *and* conscious could develop motives that might not be in the interests of humanity? This possibility has long been a concern of imaginative writers and futurists, but I would like to draw particular attention to the fact that consciousness as we have come to understand it in this book is, unlike intelligence, deeply bound up with the belief that it is 'good' to survive and reproduce. An intelligent machine whose behaviour is grounded in this value system poses special dangers, therefore, not only to human beings but also potentially to all other existing life forms. Certainly, this danger arises for any life form that is seen as a potential threat to these self machines, or even as a significant competitor for their resources. Intelligence combined with self-preservative motivation is something quite different from intelligence alone.

I will not enumerate all the ethical concerns and potential dangers that arise with the possibility of artificial sentience. There is already a large literature on this topic. It says a great deal about the current state of AI that such questions are being taken seriously by so many people in influential positions.[36] This fact alone might encourage you to view the prospect of sentient machines with more concern. Certainly, I am far more mindful of these issues than I ever was. As recently as 2017 I did not consider robot consciousness to be feasible, not only in my own lifetime but in principle. I have changed my mind about that.

So, given all these ethical concerns, why do I consider it necessary to attempt to demonstrate that consciousness can be produced artificially? Simply because this seems to be the only way that the hypotheses advanced in this book can be falsified. Unless and until we engineer consciousness, we cannot be confident that we have solved the problem as to why *and how* it arises.

Is that reason enough to take the momentous risks I have just described? My own answer to this question starts from the following belief: *If it can be done it will be done.* In other words, if it is possible in principle to engineer consciousness, then someday and somewhere it will happen.[37] This prediction applies whether the particular hypotheses advanced in this book are correct or not. My responsibilities lie with the present hypotheses, however, and with the possibility that they might be correct. If they are correct, or even on the right track, then the creation of artificial consciousness is imminent. In other words, some such hypotheses *will* soon be used to engineer consciousness.

The individual facts that led to the conclusions reported in this book – almost all of them – have been in the public domain for several years. Although it is true that many neuroscientists interpret these facts differently from me, it is also true that others have come to very similar conclusions. Although each of them has emphasised different aspects and nuanced them differently, nevertheless it is fair to say that Jaak Panksepp, Antonio Damasio and

Bjorn Merker, at least, have all come to the view that (1) conscious-ness is generated in the upper brainstem, (2) it is fundamentally affective, and (3) it is an extended form of homeostasis. These facts combined mean that consciousness is not as complicated as we previously thought. It is therefore reasonable to expect that we can engineer it. The only major addition that this book makes to these conclusions is (4) the Free Energy Principle. That, too, is not very complicated in its essence; in fact, its great appeal lies in the fact that it reduces almost all mental and neurological processes to a single mechanism, and renders them computable.

The Free Energy Principle, too, is already in the public domain. Moreover, Friston and I have already published scientific articles in which we combine this principle with the other three principles just enumerated.[38] After all, it would be unusual to publish hypotheses like these in a book aimed at a general reader-ship before subjecting them to peer review and publishing them in appropriate specialist journals. The same applies to oral presenta-tions, where one is required to defend one's claims in scientific and scholarly forums. I have presented the ideas in this book to several audiences around the world in various specialist disciplines.[39]

And so the cat is out of the bag. Does this mean that I should have held back from publishing these articles and presenting these papers? The answer is unequivocally no. If I didn't do it, some-body else would have done it. These ideas are in the air. Their time has come. I am not being defensive; this is demonstrably true. Consider for example the article by our Czech colleagues who pre-dicted in 2017 that a solution to the hard problem was imminent and that it would be grounded in the Free Energy Principle.[40] Robin Carhart-Harris (who attended some of the early neuropsychoana-lytic meetings held in London) has independently published ideas along similar lines.[41] The same applies to the social neuroscientist Katerina Fotopoulou, who recognised the link between uncer-tainty (inverse precision) and consciousness as early as 2013.[42] Undoubtedly, if the hard problem of consciousness is going to be solved through some combination of the four insights listed above, then it is destined to happen in the very near future, with or without my involvement.

This realisation guides the approach I have decided to take on the ethical and moral questions we are considering. My approach is the very opposite of holding back publications and the like; instead, it is to assist wherever possible with the practical implementation of my hypotheses – and to do so without delay. This approach flows logically from the expectation that if it can be done it will be done. I must therefore try to keep ahead of the wave, so that I am in a position to forestall the potentially harmful consequences, wherever possible.

In essence, this means that the project outlined in this chapter must be implemented *now* and it must be implemented without any commercial funding. Assuming that my research team succeeds in reaching the criterion outlined above, namely the survival of the artificial self in unpredicted environments, and assuming we obtain reasonable evidence to the effect that it is sentient (i.e. if there is no *disconfirmation* of my predictions on that score), then – in my view – we should immediately proceed with the following three steps.

First, I believe, we must switch the machine off and remove its internal battery. I realise this broaches one of the ethical concerns listed above, but I think it is the right thing to do in the first instance. We must recall that a machine of the kind envisaged here *will not be alive*. I see no reason why switching off a non-living conscious machine should entail its demise. It should always be possible to switch it on again; and presumably the conscious agent thus revived will be identical to the one that was switched off (using the biological analogy of sleep and waking). It should be noted that this plan to switch off our machine is consistent with the 'Termination Obligation' clause contained in the Universal Guidelines for Artificial Intelligence (2018), which is 'the ultimate statement of accountability for an AI system'.[43]

Second, we must begin the process of patenting the critical component of our third-stage system – its neural correlate of consciousness, whatever it turns out to be – which enabled us to reach our stated criteria. It is not possible to patent mere equations, so there is no risk that somebody else will do this while we try to implement them. However, since the equations are already in the public domain, it is imperative for us to act swiftly, so that we can control their concrete implementation before anyone else does. If it comes to this, it is important that the patent should be registered in the name of an appropriate non-profit organisation – such as OpenAI or the Future of Life Institute – rather than any individual or group of individuals. At the very least, this assures collective decision-making, and increases the chances that decisions will be made in the interests of the greater good.

Third, and finally, if our criteria are reached and the patent is registered, its custodians should organise a symposium in which leading scientists and philosophers and other stakeholders are invited to consider the implications, and to make recommendations concerning the way forward, including whether and when and under what conditions the sentient machine should be switched on again – and possibly developed further. Hopefully this will lead to the drawing up of a set of broader guidelines and constraints upon the future development, exploitation and proliferation of sentient AI in general.

I think all of this must be said, however surprised I myself am to have come to these recommendations. Having done so, we should be under no illusions as to their frailty. The precedent of nuclear energy and atomic weapons is plain for all to see. There is no alternative than to do what we can, as soon as we can, to recognise the scale of the implications arising from our imminent ability to create conscious machines.

A reasonably clear objective criterion of sentience seems now to be at hand. Hopefully, this will alter our ethical behaviour more generally, beyond the narrow questions that arise from the prospect of making conscious machines. Feelings are widely taken to be necessary and sufficient conditions for ethical concern. The scientific understanding of feelings outlined in this book therefore presents us with an opportunity to think a little more deeply about *animal* suffering. I have mentioned more than once how the advances in affective neuroscience in the late twentieth century (i.e. the realisation that what is required for sentient being is little more than a midbrain decision triangle, something that we share with all vertebrates) altered many scientists' views about what is and is not acceptable in animal research. It seems self-evident that the same should apply to the public's attitude towards animal welfare more generally. For example, how do we justify industrial-scale breeding and slaughter of fellow sentient beings for the purposes of eating them? When addressing this question, we must bear in mind that consciousness emerges by degrees, so that the putative sentience of a fly or a fish cannot be equated directly with that of a human being. By the same token, however, we must remember that sheep and cows and pigs (which feature so prominently on Western menus) are fellow *mammals*. This means they are subject to the same basic emotions that we are, such as FEAR, PANIC/ GRIEF and CARE. Mammals possess a cortex, too, which means they are capable – all of them, to some degree – of consciously 'remembering the future' and feeling their way through its probabilities and likelihoods.

As the twenty-first century unfolds, in the absence of any higher goal – if all that we are is our consciousness – what else should we do but try to minimise suffering? Now that we have a better idea of where suffering might exist, what else could we do with this knowledge? The preservation and protection of biological consciousness is decidedly not tied to the fate of our species alone.

Considering all that I have asked you to relinquish through the pages of this book, with regard to human exceptionalism and the like, it may be fitting to end with a brief reflection upon what we might gain in our self-conception from these unwelcome insights.

Feeling is a precious inheritance. It carries within it the wisdom of the ages: an inheritance that extends backwards over aeons to the beginning of life itself. When homeostasis eventually gave rise to feelings, the crux of this new capacity was that it enabled us to know *how we are doing* within a biological scale of values. Feelings entrain predictions that are grounded in the accumulated experiences in situations of biological significance of literally all our ancestors. Feelings enable us to do what is best for us, even as we do not know *why* we do so. I have asked you before to imagine what would happen if each of us had to learn afresh which foodstuffs contain high energy supplies and if we had to discover for ourselves what happens when we jump off cliffs. Due to the unbidden feelings that attract us to sweetness and make us avoid heights, we 'just know' (at a first approximation) what to do and when. For example, we know what to do when babies cry, predators attack, or frustrating obstacles get in our way. This innate knowledge – which is conveyed to us explicitly *only* in the form of feelings – is what makes it possible for us to survive in the highly unpredictable worlds that we do, where motor vehicles hurtle around us and carbon dioxide fills the air.

So, as we relinquish the familiar illusion that consciousness flows in through our senses, and the misconception that it is synonymous with understanding, let us take comfort in the fact that it actually comes spontaneously from our inmost interior. It dawns within us even before we are born. At its source, we are guided by a constant stream of feelings, flowing from a wellspring of intuition, arising from we know not where. Each of us individually does not know the causes, but we feel them. Feelings are a legacy that the whole history of life has bestowed upon us, to steel us for the uncertainties to come.

Postscript

Shortly after writing a full draft of this book, I was invited to present its main thesis at the annual Science of Consciousness congress held at Interlaken in 2019. This required me to distil most of what you have just read into the format of a plenary lecture. It might be useful to end our lengthy journey by summarising the thirteen points I used for that lecture.

(1) The great nineteenth-century physiologist Johannes Müller believed that animate organisms 'contain some non-physical element or are governed by different principles than are inanimate things'. His students (Helmholtz, Brücke, Du Bois-Reymond, Ludwig and others) disagreed; they were certain that 'no other forces than the common physical and chemical ones are active within the organism'. Their pupil, Sigmund Freud, in turn, tried to establish a natural science of the mind on this basis, in which mental life could be reduced to 'quantitatively determinate states of specifiable material particles'. He failed in his project, lacking the methods, and abandoned it in 1896.

(2) A century later (1994), the pioneering biologist Francis Crick declared that 'you, your joys and your sorrows, your memories and your ambitions, your sense of personal identity and free will, are in fact no more than the behaviour of a vast assembly of nerve cells and their associated molecules'. He exhorted us to try again to discover the neural correlates of consciousness and he attempted to do so himself. Unfortunately, however, he used visual consciousness as his model example.

(3) In response, the philosopher David Chalmers argued that Crick's search for the neural correlates of consciousness was an 'easy' problem – a correlational rather than a causal one – the solution of which could explain where but not why and how consciousness arises. For Chalmers, the 'hard' problem of consciousness was: how and why do neurophysiological activities produce the experience of consciousness? For him (and his philosophical predecessor Thomas Nagel) the problem revolved around the something-it-is-like-ness of experience: 'An organism has conscious mental states if and only if there is something that it is like to *be* that organism – something it is like *for* the organism.' The hard problem, therefore, is this: why and how does the subjective quality of experience arise from objective neurophysiological events?

(4) To ask how objective things *produce* subjective things is to speak loosely; it risks making the hard problem harder than it needs to be. Objectivity and subjectivity are observational perspectives, not causes and effects. Neurophysiological events can no more produce psychological events than lightning can produce thunder. They are parallel manifestations of a single underlying process. The underlying cause of both lightning and thunder is electricity, the lawful mechanisms of which explain them both. Physiological and psychological phenomena can likewise be reduced to unitary causes, but not to each other.

(5) We usually describe the underlying causes of biological phenomena in 'functional' terms, and functional mechanisms can in turn be reduced to natural laws. For example: what is the mechanism of vision? However, Chalmers correctly points out that the functional mechanism of vision does not explain what it is like to see. This is because vision is not an intrinsically conscious function. The performance of visual functions (even specifically human ones, like reading) need not feel like anything. Perception readily occurs without awareness of what is perceived, and

learning without awareness of what is learnt. Therefore, Chalmers reasonably asked: 'Why is the performance of these functions accompanied by experience? Why doesn't all this information processing go on "in the dark", free of any inner feel?' Science's failure to answer this question raises the possibility that consciousness does not form part of the ordinary causal matrix of the universe.

(6) Chalmers's question may reasonably be asked of all cognitive functions, not only visual ones, but the same does not apply to affective functions. How can you have a feeling without feeling it? How can we explain the functional mechanism of affect without explaining why and how it causes us to experience something? Even Freud agreed on this score: 'It is surely of the essence of an emotion that we should be aware of it, i.e. that it should become known to consciousness. Thus, the possibility of the attribute of unconsciousness would be completely excluded as far as emotions, feelings, and affects are concerned.'

(7) Against this background, it is of the utmost interest to observe that cortical functioning is accompanied by consciousness only if it is 'enabled' by the reticular activating system of the upper brainstem. Damage to just two cubic millimetres of this region obliterates all consciousness. Many people believe that this is because the brainstem modulates the quantitative level of consciousness, or 'wakefulness'; but that view is unsustainable. The consciousness generated by the upper brainstem has qualitative content of its own. This is affect. Since cortical consciousness is contingent upon brainstem consciousness, affect is revealed to be the foundational form of consciousness. The sentient subject is literally constituted by affect.

(8) Affect is an extended form of homeostasis, which is a basic biological mechanism that arose naturally with self-organisation.

Self-organising systems survive because they occupy limited states; they do not disperse themselves. This survival imperative led gradually to the evolution of the complex dynamical mechanisms that underwrite intentionality. Crucially, the selfhood of self-organising systems grants them a point of view. That is why it becomes meaningful to speak of the subjectivity of such a system: deviations from their viable states are registered by the system, for the system, as needs.

(9) Affect hedonically valences biological needs, so that increasing and decreasing deviations from homeostatic settling points (increasing and decreasing prediction errors) are felt as unpleasure and pleasure respectively. Each category of need – of which there is a great variety – has an affective quality of its own and each triggers action programmes which are predicted to return the organism to its viable bounds. These active states – i.e. intentional responses to the affective states – take the form of innate reflexes and instincts, which are gradually supplemented by learning from experience in accordance with the Law of Affect.[1] Feeling by an organism of fluctuations in its own needs enables choice and thereby supports survival in unpredicted contexts. This is the biological function of experience.

(10) Needs cannot all be felt at once. They are prioritised by a midbrain decision triangle, where current needs (residual prediction errors, quantified as free energy) converging on the periaqueductal grey are ranked in relation to current opportunities (displayed in the form of a two-dimensional 'saliency map' in the superior colliculi). This triggers conditioned action programmes, which unfold in expected contexts over a deep hierarchy of predictions (the generative model of the expanded forebrain). The actions that are generated by prioritised affects are voluntary, which means they are subject to here-and-now choices rather than pre-established algorithms. Such choices are felt in exteroceptive

consciousness, which contextualises affect. The choices are made on the basis of fluctuating precision-weighting (a.k.a. arousal, modulation, post-synaptic gain) of the incoming error signals that are rendered salient by prioritised needs, while they are buffered in working memory, with the aim of minimising uncertainty (maximising confidence) in a current prediction as to how the need can be met. This is 'reconsolidation'. As Freud said, 'consciousness arises instead of a memory trace'.

(11) Reliably successful choices result in long-term adjustments of sensory-motor predictions. Thus, exteroceptive consciousness is predictive work in progress, the aim of which is to establish ever deeper (more certain, less conscious) predictions as to how needs may be resolved. This long-term consolidation – and the transition from 'declarative' to 'non-declarative' memory systems – requires reduction of complexity in the predictive model, to facilitate generalisability. We aspire to automaticity – absolute confidence – but we can never achieve it completely. To the extent that we fail, we suffer feelings. Since we never achieve errorless prediction, the default drive (when all goes well) is SEEKING – proactive engagement with uncertainty, with the aim of resolving it in advance. When this affect is prioritised, it is felt as curiosity and interest in the world.

(12) These are the causal mechanisms of consciousness – in both its manifestations, neurological and psychological – what it looks like and what it feels like. The underlying functions can be reduced to natural laws, such as Friston's Law.[2] These laws underwrite self-organisation. They are no less capable of explaining how and why proactively resisting entropy (i.e. oblivion) feels like something than other scientific laws are capable of explaining other natural things. Consciousness is part of nature and it is mathematically tractable.

(13) All known conscious systems are alive but not all living systems are conscious. Likewise, all living systems are self-evidencing but not all self-evidencing systems are alive. If the argument laid out here is correct, then, in principle, an artificially conscious self-evidencing system can be engineered. Consciousness can be *produced*. This will realise the wildest dreams of Helmholtz and other members of the Berlin Physical Society. However, we must question our motives for doing this, accept collective responsibility for the potentially dire consequences and proceed with extreme caution.

Appendix: Arousal and Information

In an authoritative book on the topic of brain arousal, Pfaff (2005, pp. 2–6) comments as follows:

> Satisfying the need for an 'energy source' for behavior, arousal explains the initiation and persistence of motivated behavior in a wide variety of species [...] Arousal, fuelling drive mechanisms, potentiates behavior, while specific motives and incentives explain why an animal does one thing and not another [...] The *Dictionary of Ethology* not only emphasizes arousal in the context of the sleep–wake cycle but also refers to the overall state of responsiveness of the animal, as indicated by the intensity of stimulation necessary to trigger a behavioral reaction. Arousal 'moves the animal towards readiness for action from a state of inactivity.' In the case of directed action, a founder of ethology, Niko Tinbergen, would say arousal provides the motoric energy for a 'fixed action pattern' in response to a 'sign stimulus.' The dictionary does not eschew neurophysiology, as it also covers arousal levels indicated by the cortical electroencephalogram (EEG) [...] Generations of behavioral scientists have both theorized and experimentally confirmed that a concept like arousal is necessary to explain the initiation, strength, and persistence of behavioral responses. Arousal provides the fundamental force that makes animals and humans active and responsive so they will perform instinctive behaviors or learned behaviors directed toward goal objects. The strength of a learned response depends on arousal and drive. Hebb

saw a state of generalized activation as fundamental to optimal cognitive performance. Duffy goes even further by invoking the concept of 'activation' to account for a significant part of an animal's behavior.

Pfaff's own principal component analyses suggest that the proportion of behaviour across a wide range of data that can be accounted for by 'generalised arousal' is between 30 per cent and 45 per cent.

[Duffy] anticipated that quantitative physiologic or physical measures would allow a mathematical approach to this aspect of behavioral science [...] Cannon brought in the autonomic nervous system as a necessary mechanism by which arousal prepares the animal for muscular action. Entire theories of emotion were based on the activation of behavior [...] Malmo brought all of this material together by citing EEG evidence and physiologic data, which go along with behavioral results in establishing activation and arousal as primary components driving all behavioral mechanisms [...] This is the classic arousal problem: How do internal and external influences wake up brain and behavior, whether in humans or in other animals, whether in the laboratory or in natural, ethological settings? It is important to reformulate and solve this problem because we are dealing with responsivity to the environment, one of the elementary requirements for animal life. It is also timely to reformulate and solve this problem now because new neurobiologic, genetic, and computational tools have opened up approaches to 'behavioral states' that were never possible before [...] Explaining arousal will permit us to understand the states of behavior that lie beneath large numbers of specific response mechanisms. Not only is it strategic to accomplish the analysis of many behaviors all at once but also elucidating mechanisms of behavioral states leads to an understanding of mood and temperament. To put it another way, much of twentieth-century neuroscience was directed at explaining the particularity of specific stimulus/response connections. Now we are in a position to reveal mechanisms of entire classes of responses under the name of 'state control'. Most important are the mechanisms determining the level of arousal [...] Any truly universal definition of arousal must be elementary and fundamental, primitive and undifferentiated, and not derived from higher CNS functions. It cannot be limited by particular, temporary conditions or measures. For example, it cannot be confined to explaining responses to only one stimulus modality. Voluntary motor activity and emotional responses should also be included. Therefore, I propose the following as an operational definition that is intuitively satisfying and that will lead to precise quantitative measurements: *'Generalized arousal' is higher in an animal or human being who is: (S) more alert to sensory stimuli of all sorts, and (M) more motorically active, and (E) more reactive emotionally.* This is a concrete definition of the most fundamental force in the nervous system [...] All three components can be measured with precision [...] Clearly there is a neuroanatomy of generalized arousal, there are neurons whose firing patterns lead to it, and genes whose loss disrupts it. Therefore [...] generalized arousal is the behavioral state produced by arousal pathways, their electrophysiological mechanisms, and genetic influences. The fact that these mechanisms produce the same sensory alertness (S), motor reactivity (M) and emotional reactivity (E) as our definition affirms the existence of a generalized arousal function and the accuracy of its operational definition.

Pfaff continues: 'Because CNS arousal depends on surprise and unpredictability, its appropriate quantification depends on the mathematics of *information*' (p. 13, emphasis added). Shannon's (1948) equation makes information measurable, as Pfaff explains:

> If any event is perfectly regular, say the ticking of a metronome, the next event (the next tick) does not tell us anything new. It has an extremely high probability (p) of occurrence in exactly that time bin [...] We have no uncertainty about whether, in any given time bin, the tick will occur. In Shannon's equation, the information in any event is in inverse proportion to its probability. Put another way, the more uncertain we are about the occurrence of that event, the more information is transmitted, inherently, when it does happen [...] When all events in an array of events are equally probable, information is at its top value. Disorder maximizes information flow. Coming from thermodynamics, the technical term for disorder in Shannon's equation is entropy. His symbol for entropy is H [...] The information content inherent in some event x is:

$$H(x) = p(x) \log_2 \frac{1}{p(x)}$$

where $p(x)$ is the probability of event x.

Pfaff sums up (pp. 19–20):

For a lower animal or human to be aroused, there must be some change in the [interoceptive or exteroceptive] environment. If there is change, there must be some uncertainty about the state of the environment. Quantitatively, to the degree that there is uncertainty, predictability is decreased. Given these considerations, we can use [Shannon's equation] to state that the less predictable the environment and the greater the entropy, the more information is available. Arousal of brain and behavior, and information calculations, are inseparably united.

In short, unknown, unexpected, disordered and unusual (high-information) stimuli produce and sustain arousal responses (p. 23).

Information theory has been lurking behind behavioral investigations and neurophysiologic data all along. First, in clear and simple logic, consider what is required for an animal or human being to rouse itself to action. Second, consider what is required to recognize a familiar stimulus (habituation) and to give special attention to a novel stimulus. Third, from the experimenter's point of view, information theory provides methods for calculating the meaningful content of spike trains and quantifying the cognitive load of certain environmental situations. New questions can be asked: How much distortion of a sensory stimulus field is required for novelty? What kinds of generalization from a specific type of stimulus are allowed for a given type of response? The information theoretic approach will help us to turn the combination of genetics, neurophysiology, and behavior into a quantitative science. We can use the 'mathematics of arousal' to help analyse neurobiologic mechanisms.

Pfaff ultimately concludes (pp. 138–45):

CNS arousal systems battle heroically against the Second Law of Thermodynamics in a very special way. They respond selectively to environmental situations that have an inherently high entropy – a high degree of uncertainty and therefore information content. But in responding, CNS arousal systems effectively reduce entropy by compressing all of that information into a single, lawful response [...] Arousal neurobiology is the neuroscience of change, uncertainty, unpredictability, and surprise – that is, of information science. Throughout all of the analyses of arousal mechanisms in the CNS so far – neuroanatomic, physiologic, genetic, and behavioural – the concepts of information theory have proven useful. The mathematics of information provides ways of classifying responses to natural stimuli. Nerve cells actually encode probabilities and uncertainties, with the result that they can guide behavior in unpredictable circumstances. CNS arousal itself absolutely depends on change, uncertainty, unpredictability, and surprise. The huge phenomenon called habituation, a decline in response amplitude on repetition of the same stimulus, pervades neurophysiology, behavioral science, and autonomic physiology; and it shows us how declining information content leads to declining CNS arousal. Thus, arousal theory and information theory were made for each other.

It is important to recognise that the 'mathematics of information' explains the behaviour of neurons in both arousal processes and learning processes, which, combined, determine what the brain *does*. Therefore, although 'information' is not a physiological construct, it lawfully explains the physiological activity of the brain. It is the *function* that is selected by evolution; the physiological phenotypes follow.

Acknowledgements

I am grateful to the following friends and colleagues for reading successive drafts of the chapters of this book: Richard Astor, Nikolai Axmacher, Samantha Brooks, Aimee Dollman, George Ellis, Karl Friston, Eliza Kentridge (who is much more than a friend), Joe Krikler, Joshua Martin, Lois Oppenheim, Jonathan Shock, Pippa Skotnes and Dawie van den Heever. I am especially indebted to Ed Lake for making the manuscript so much more readable; I have never seen an editor work so hard. And thanks to my agent, Caroline Dawnay, without whom Ed wouldn't have known of it. The manuscript was copy-edited by Trevor Horwood, with collateral input by Tim James.

I would like also to thank Sir Sydney Kentridge for giving me the use of his Chailey house, where the bulk of this book was written over the winters of 2018/19 and 2019/20. Behind the scenes, as always, were my intrepid assistants, Paula Barkay and Eleni Pantelis. This book, like most else I have done, would not have happened without them.

Notes

Chapter 1: The Stuff of Dreams

1. Popper (1963). It is true that not everyone agrees with this formulation of how science works. Nevertheless, it is the one that almost all natural scientists endorse. *Now that you are reading these endnotes, an explanation is in order. They are aimed mainly at academic readers who are interested in (or have a background in) the various technical literatures that this book trenches on. The notes can safely be ignored by general readers, who are my primary audience.*

2. Freud (1893a), p. 13.

3. Sacks (1984), p. 164.

4. Freud (1895), p. 160. Some thirty-six years later, Freud wrote movingly to Albert Einstein about the lack of scientific standing of psychology compared with physics (Freud, 1994, p. 239):

    Admittedly, it is not altogether a matter of regret that one has opted for psychology. There is no greater, richer, more mysterious subject, worthy of every effort of the human intellect, than the life of the mind. Psychology surely is the most lovely of all the noble ladies; it is just that her knight is doomed to remain unhappy in his love.

5. 'I *love* the passage from Freud – I am very happy you tracked it down. As you say, generously, something analogous might be said of my own case-histories, and of neurological (at least neuropsychological) case-histories in general. I have quoted it (tho' I don't know whether it will survive – my manuscript has become far too long and footnotey) in a just-completed, rather general piece – "Scotoma" – about forgetting and neglect in science' (letter from Sacks dated 2 January 1995).

6. As the neuroscientist Semir Zeki wrote at the time: 'Most [of us] would shrink in horror at the thought of investigating what appears so impenetrable a problem' (Zeki, 1993, p. 343).

7. Aserinsky and Kleitman (1953).

8. Dement and Kleitman (1957).

9. See Freud (1912), p. 264–65:

   There is one psychical product to be met with in the most normal persons, which yet presents a very striking analogy to the wildest productions of insanity, and was no more intelligible to philosophers than insanity itself. I refer to dreams. Psychoanalysis is founded upon the analysis of dreams; the interpretation of dreams is the most complete piece of work the young science has done up to the present.

10. Popper (1963).

11. From the strictly anatomical viewpoint, the thalamus is not considered part of the brainstem. Physiologically, however, some of its 'non-specific' nuclei are part of the reticular activating system; they are therefore grouped together with the functions of the brainstem (hence the term 'extended reticulo-thalamic activating system', ERTAS). The 'specific' thalamic nuclei, which act mainly as relay stations for sensory signals, are grouped together with the functions of the cortex. In this book, I will use the terms 'brainstem' and 'cortex' mainly to designate the functional-anatomical division between ERTAS *arousal* and thalamocortical *representation* respectively. I will therefore count not only the 'non-specific' thalamus and hypothalamus as brainstem structures, but the basal forebrain nuclei too. For a contemporary view of the *extended* reticulo-thalamic activating system, see Edlow et al. (2012).

12. Jouvet (1965).

13. Hobson, McCarley and Wyzinski (1975).

14. McCarley and Hobson (1977), p. 1346.

15. McCarley and Hobson (1977), p. 1219.

16. The one and only patient who did report loss of dreaming almost certainly sustained damage well beyond the REM-generating mesopontine tegmentum (due to traumatic subarachnoid haemorrhage; Lavie et al., 1984). Changes in his dreaming were therefore difficult to link with any particular brain region.

17. See Solms (1997a) for a full description of my findings. My thesis was submitted in 1991, but I only got around to publishing it six years later.

18. Solms (2000a). The principle of 'double dissociation' in neuropsychology enables us to carve mental functions at their joints: if damage in area X (of the brain) causes loss of function A but not function B, and damage in area Y causes loss of function B but not function A, then function A and B cannot be the same thing. In other words, in this case, the function of REM sleep and the function of dreaming cannot be the same. They

are *correlated* with each other (i.e. they occur at the same time) but they are not the same thing.

19. Many other facts, in addition to my lesion findings, support this conclusion. For example, there is a 50 per cent chance of obtaining dream reports during the first few minutes of sleep (in descending Stage 2), long before the first REM episode. Likewise, dreams which are completely indistinguishable from REM ones occur in non-REM sleep with increasing frequency during the rising morning phase of the diurnal rhythm. This is called the 'late-morning effect'. Also, although dreaming is much more frequent in REM than non-REM sleep, non-REM sleep is more prevalent than REM sleep. Consequently, at least a quarter of all dreams occur in non-REM sleep. See Solms (2000a) for details.

20. Solms (1991, 1995).

21. Frank (1946, 1950), Partridge (1950).

22. Schindler (1953).

23. Hartmann et al. (1980).

24. Sharf et al. (1978). Subsequent studies showed that dopamine antagonists have the opposite effect (Yu, 2007).

25. Dahan et al. (2007).

26. Léna et al. (2005).

27. Solms (2011).

28. Solms (2001).

29. This is a bad name; a throwback to behaviourist times. There are many different varieties of 'reward' (i.e. pleasure) in the brain.

30. Rolls (2014), Berridge (2003), Panksepp (1998).

31. Panksepp (1998), p. 155.

32. Dopaminergic SEEKING activity (unlike other monoamine activity) continues with sleep onset, and it is maximal during REM sleep. It is perhaps not coincidental that this coincides with rapid saccadic eye moments. Eye movements in humans, like sniffing and whisker twitching in rodents, are a good proxy for SEEKING activation (see Panksepp, 1998).

33. Pace-Schott and Hobson (1998).

34. Braun (1999), p. 196.

35. Ibid., p. 201.

Chapter 2: Before and After Freud

1. I was accepted for the training then but didn't actually begin it until 1989.

2. Actually, Freud did not give this unpublished manuscript a title; the title was invented by its English translators. In his correspondence with Wilhelm Fliess, Freud called it a 'Psychology for Neurologists', 'Sketch of a Psychology' and 'the Psychology'.

3. Letter to Hallmann (1842), published in Du Bois-Reymond (1918), p. 108. Also often quoted is Du Bois-Reymond's Preface to his *Über die Lebenskraft* (1848–84, pp. xliii–iv):

    There are no new forces in operation in organisms and its particles, no forces that are not also in operation outside of them. There are also no forces which deserve the name of 'vital force'. The separation between the so-called organic and inorganic natures is completely arbitrary.

4. Bechtel and Richardson (1998).

5. See my discussion of this term in 'Solms (2021b)'.

6. Freud (1950b), p. 295.

7. Freud's priority in formulating the 'functionalist' position is not widely recognised (Freud, 1900, p. 536):

    [We] attempt to make the complications of mental functioning intelligible by dissecting the function and assigning its different constituents to different component parts of the apparatus. So far as I know, the experiment has not hitherto been made of using this method of dissection in order to investigate the way in which the mental instrument is put together, and I can see no harm in it.

    Cf. Shallice (1988).

8. When Freud first introduced this curious term, he explained that it refers to a level of explanation which incorporates both psychology and biology (letter to Fliess of 10 March 1898; Freud, 1950a). As Freud once wrote to Georg Groddeck: 'The unconscious is the long-sought missing link between the physical and the mental' (letter of 5 June 1917). See Solms (2000b). See also my presentation to a meeting of the New York Academy of Sciences held to commemorate the centenary of Freud's 'Project' (Solms, 1998). The great Karl Pribram spoke at that meeting. I also met the pioneering neurophysiologist, Joseph Bogen, at that conference. I vividly recall him casually saying that consciousness was generated by the intralaminar nuclei of the thalamus, which was the first time I heard anyone suggest that cortex was not intrinsically conscious. See Bogen (1995).

9. Freud's (1891) critique of localisationism laid the foundations for the 'functional systems' approach that dominated later neuropsychology and, subsequently, cognitivism. I discussed this issue in detail in Solms and Saling (1986) and Solms (2000b).

10. Hence the title of Sulloway's (1979) book, *Freud: Biologist of the Mind*.

11. Freud (1914), p. 78.

12. Freud (1920), p. 60.

13. Letter to Fliess of 20 October 1895.

14. Letter to Fliess of 29 November 1895.

15. Letter to Fliess of 25 May 1895.

16. I am currently preparing an English translation of *The Complete Neuroscientific Works of Sigmund Freud* (in 4 volumes). See also my revision of Strachey's translations: *The Revised Standard Edition of the Complete Psychological Works of Sigmund Freud* (24 volumes).

17. Freud (1950b), pp. 303, 316. Many years before Freud, Baruch Spinoza wrote: 'desire is the very nature or essence of a person'.

18. Freud (1901), p. 259.

19. Freud (1920), p. 60.

20. Freud (1915a), pp. 121–2, emphasis added.

21. Letter to Fliess of 25 May 1895.

22. Freud (1940), p. 197.

23. This figure is adapted from Braun et al. (1997). Braun's study was purely descriptive. His findings are compatible with Freud's theory but they do not confirm it experimentally because they did not test any predictions derived from it. However, a student of mine (Catherine Cameron-Dow, 2012) recently tested directly Freud's theory to the effect that dreams protect sleep. Her confirmation of the hypothesis is currently being followed up in a larger study by my colleague Tamara Fischmann in Berlin.

24. Incidentally, the debate was chaired by none other than David Chalmers. The result reversed a 1977 vote that had followed a presentation by Hobson of his 'activation-synthesis' theory to the assembled members of the American Psychiatric Association.

25. Solms and Saling (1986).

26. See Braun (1999).

27. Malcolm-Smith et al. (2012).

28. Solms (2000c), Solms and Zellner (2012).

29. Interestingly, Freud developed his 'libidinal drive' concept at a time that he was regularly using cocaine, an alkaloid which powerfully activates

the dopaminergic SEEKING system. Is it too fanciful to suppose that Freud's personal experience of the generalised motivational effects of cocaine contributed to his recognition of the existence of such an all-purpose motivational mechanism in the mind?

30. In Chapter 7, I will link 'wish-fulfilment' with predictive coding and 'reality testing' with what is nowadays called 'prediction error' (or precision-modulated prediction error).

31. Fotopoulou, Solms and Turnbull (2004).

32. Turnbull, Jenkins and Rowley (2004).

33. Fotopoulou and Conway (2004), Turnbull, Berry and Evans (2004), Fotopoulou et al. (2007, 2008a,b), Turnbull and Solms (2007), Fotopoulou, Conway and Solms (2007), Fotopoulou (2008, 2009, 2010a,b), Coltheart and Turner (2009), Cole et al. (2014), Besharati, Fotopoulou and Kopelman (2014), Kopelman, Bajo and Fotopoulou (2015).

34. See Turnbull, Fotopoulou and Solms (2014) for review. See also Besharati et al. (2014, 2016).

35. Zellner et al. (2011).

36. Solms and Turnbull (2002, 2011), Panksepp and Solms (2012), Solms (2015a).

37. These presentations were eventually collected together in volume form: Kaplan-Solms and Solms (2000).

38. Kandel (1998).

39. Kandel (1999), p. 505.

40. The neuroscientists included Allen Braun, Jason Brown, Antonio Damasio, Vittorio Gallese, Nicholas Humphrey, Eric Kandel, Marcel Kinsbourne, Joseph LeDoux, Rodolfo Llinás, Georg Northoff, Jaak Panksepp, Michael Posner, Vilanayur Ramachandran, Oliver Sacks, Todd Sacktor, Daniel Schacter, Carlo Semenza, Tim Shallice, Wolf Singer and Max Velmans. The psychoanalysts included Peter Fonagy, Andre Green, Ilse Grubrich-Simitis, Otto Kernberg, Marianne Leuzinger-Bohleber, Arnold Modell, Barry Opatow, Allan Schore, Theodore Shapiro, Riccardo Steiner and Daniel Widlöcher.

41. Damasio, Damasio and Tranel (2013).

42. Freud (1940), p. 198. Throughout this book, I am using my revised versions of James Strachey's translations (see Solms, 2021b).

43. Freud (1920), p. 24, emphasis added.

44. Freud (1940), pp. 161–2, emphasis added.

45. See Freud (1923), p. 26:

The ego is first and foremost a bodily ego; it is not merely a surface entity, but is itself the projection of a surface. If we wish to find an anatomical analogy for it we can best identify it with the 'cortical homunculus' of the anatomists, which stands on its head in the cortex, sticks up its heels, faces backwards and, as we know, has its speech-area on the left-hand side [...] The ego is ultimately derived from bodily sensations, chiefly from those springing from the surface of the body. It may thus be regarded as a mental projection of the surface of the body.

46. See Solms (2013).

Chapter 3: The Cortical Fallacy

1. Merker (2007), p. 79. These observations confirmed an earlier report by Shewmon, Holmes and Byrne (1999).
2. The following account of Merker's findings is paraphrased from his (2007) published report.
3. Damasio and Carvalho (2013), p. 147.
4. These observations paraphrase Merker's summary of the literature (2007), p. 74.
5. Panksepp (1998).
6. See Weiskrantz (2009).
7. Watch the video: https://blogs.scientificamerican.com/observations/blindsight-seeing-without-knowing-it/.
8. The same applies to the pathway from the eye to the lateral geniculate body. See Figure 6.
9. Zeman (2001).
10. Coenen (2007), p. 88, emphasis added.
11. No *eye opening* = 1 point, opening in response to pain = 2 points, opening in response to speech = 3, opening spontaneously = 4; no *verbal response* = 1, response with incomprehensible sounds = 2, response with inappropriate words = 3, coherent but inappropriate response = 4, appropriate response = 5; no *motor response* = 1, decerebrate posturing = 2, decorticate posturing = 3, withdrawal from pain = 4, localising to pain = 5, obeying commands = 6.
12. It has also been argued that the illusion arises from the 'moralistic fallacy'. The cognitive neuroscientist Heather Berlin described it as 'arbitrary' to assume that outer responsiveness implies inner consciousness. She elaborated (Berlin, 2013, pp. 25–6):

> Solms's primary assumption that hydranencephalic children are conscious is unwarranted. We cannot assume that having a sleep–wake cycle and

expressions of emotion (laughter, rage, etc.) necessitates consciousness [...] While it is true that they may in fact be conscious, we cannot assume that they are. Unconscious processes can be quite sophisticated and complex (Berlin, 2011). The crux of Solms's theory relies on a projection of the existence of consciousness based on what look like meaningful emotional behaviors, an example of the 'moralistic fallacy' (arguing that something must be true because it would make us feel good to believe it). Humans have a natural desire to assume that consciousness exists.

This is what I wrote in response (Solms, 2013, pp. 80–81):

Why should we assume that contextually appropriate emotional displays, which are readily evoked by stimulation of a particular brain region and obliterated by lesions of that same brain region and which correspond to affective feelings in ourselves, do *not* correspond to affective feelings in these children and animals? Surely that assumption would be more 'arbitrary' than mine. The only evidence for it is that these children and animals cannot 'declare' their feelings *in words*.

13. Known as the Laws of Association: by contiguity, repetition, attention, similarity, etc.

14. Apperception is 'the process by which new experience is assimilated to and transformed by the residuum of past experience of an individual to form a new whole' (Runes, 1972).

15. Meynert (1867).

16. This conveniently overlooks the fact that it is larger in some other mammals (such as elephants). The human cortex is not even bigger than that of some other mammals in terms of the ratio between cortex and body size, or between cortex and subcortex.

17. Campbell (1904), pp. 651–2.

18. Meynert (1884; English trans. 1885), p. 160.

19. Munk (1878, 1881).

20. The anatomical basis for the distinction between blindness and mind-blindness was thought to be the fact, discovered by Paul Flechsig (1901, 1905), that striate 'projection' cortex contains primordial cells which are connected directly to the retinal periphery. These primordial cells are myelinated at birth, and therefore do not contain *memory* images. The surrounding 'association' cortex – the vehicle of all mental functions – is myelinated much later. The same applies to the other modality-specific cortices.

21. Wilbrand (1887, 1892). See my English translation of Wilbrand's original

case report, where many of these theoretical points are discussed in detail (Solms, Kaplan-Solms and Brown, 1996).

22. This observation of Wilbrand's built upon an earlier, similar observation by Charcot (1883) of a patient who experienced *non-visual* dreams, thus giving rise to the concept of 'Charcot-Wilbrand syndrome': inability to revisualise and recognise visual objects by day and loss of dreaming by night. See Solms, Kaplan-Solms and Brown (1996) and Solms (1997a) for a critical discussion of the Charcot-Wilbrand concept in light of my subsequent findings.

23. Accordingly, he drew a distinction between common deafness and 'word-deafness' (aphasia). Wernicke's aphasia was said to be caused by damage to the cortical area containing auditory memory images for words – memories of speech sounds – whereas common deafness arose from damage to the subcortical pathways connecting this area to incoming auditory sensations. Paul Pierre Broca (1861, 1865) had earlier described a parallel form of aphasia, caused by damage to the motor images for words – that is, damage to learnt programmes of how to *produce* speech sounds. According to Wernicke's student, Ludwig Lichtheim (1885), yet other forms of aphasia arose from damage to the transcortical 'association' pathways leading from the auditory and motor memory images for words to *abstract* ideas, which gave the concrete images their meaning.

    Heinrich Lissauer (1890) similarly subdivided 'mind-blindness' into two types, an 'apperceptive' type caused by damage to the visual memory images themselves and an 'associative' type caused by damage to the transcortical pathways leading from the memory images for visual objects to those for abstract ideas. It was Freud (1891) who later renamed mind-blindness as visual 'agnosia'.

    Hugo Liepmann (1900) then likewise divided 'psychical paralysis' (apraxia) into apperceptive and associative types – 'limb-kinetic' and 'ideomotor' respectively – as mentioned in the text.

24. Liepmann (1900).

25. In the Preface, he wrote the following: 'The reader will find no other definition of 'Psychiatry' in this book but the one given on the title page: Clinical Treatise on Diseases of the Forebrain. The historical term for psychiatry, i.e., "treatment of the soul", implies more than we can accomplish, and transcends the bounds of accurate scientific investigation.'

26. See Absher and Benson (1993) and Goodglass (1986). Interestingly, Antonio Damasio was a student of Geschwind's.
27. Incidentally, we know that cortical patients without language retain full consciousness, because they can and do communicate their feelings in other ways. See Kaplan-Solms and Solms (2000) for detailed descriptions of non-verbal introspective reports by aphasic patients of various types. After some initial hesitation in the nineteenth century, the general consensus has always been that loss of language does not affect 'intelligence' in any significant way.
28. Merker (2007), p. 65.
29. Cf. Ned Block's (1995) distinction between 'phenomenal' and 'access' consciousness.
30. See Craig (2009, 2011).
31. See Dehaene and Changeux (2005); Baars (1988, 1997). The term 'higher-order thought' is from Rosenthal (2005).
32. Qin et al. (2010), Mulert et al. (2005).
33. The dialogue below is extracted from an oral report of the case that was presented at the 2011 Neuropsychoanalysis Congress in Berlin. The case study was subsequently published in Damasio, Damasio and Tranel (2013).
34. Damasio, Damasio and Tranel (2013).
35. Harlow (1868).
36. LeDoux and Brown (2017).
37. See LeDoux (1999), p. 46, emphasis added:
    When electrical stimuli applied to the amygdala of humans elicit feelings of fear (see Gloor, 1992), it is not because the amygdala 'feels' fear, but instead because the various networks that the amygdala activates ultimately provide working memory with inputs that are *labelled* as fear.
38. Whitty and Lewin (1957), p. 73.
39. Solms (1997a), p. 186, Case 22.
40. The epithet was introduced by Karl Jaspers (1963).

Chapter 4: What is Experienced?

1. See Solms and Saling (1990) for a detailed discussion of Freud's disagreements with Meynert and other neuropsychological contemporaries. He first set out his dissenting views in an unpublished book manuscript (Freud 1887). They first appeared in print in Freud (1888). He then reiterated them in his (1891) book and developed them

further in Freud (1893b), and finally in his letter to Fliess of 6 December 1896.

2. Freud (1891), my translation. Freud (1886), p. 14, reported that he made 'repeated visits' to Munk's laboratory in Berlin. The course of neural pathways in the brainstem was a major focus of Freud's own anatomical research in the 1880s.

3. Freud (1891), my translation.

4. Ibid., my translation. Freud also questioned the anatomical and physiological basis of the distinction that Meynert and others drew between the apperceptive and associative stages of cortical processing (ibid.).

5. Letter to Wilhelm Fliess dated 6 December 1896 (Freud, 1950a, p. 233).

6. Freud's unconscious and preconscious systems therefore coincide with what we nowadays call 'non-declarative' and 'declarative' long-term memory, and his conscious system coincides with what we call short-term memory. The first and last of Freud's five stages (perception and consciousness) are *states* of neurons rather than traces.

7. Kihlstrom (1996).

8. Bargh and Chartrand (1999), p. 476.

9. Squire (2009).

10. Claparède (1911).

11. Galin (1974), p. 573.

12. See McKeever (1986).

13. Crick and Koch (1990), Newman and Baars (1993), Dehaene and Naccache (2001), Bogen (1995), Edelman (1990), Marc and Llinas (1994), Tononi (2012).

14. Nagel (1974). See also Chalmers (1995a) and Strawson (2006).

15. Freud (1915b), p. 177, explained:

> The whole difference [between ideas and affects] arises from the fact that ideas are cathexes – basically of memory-traces – while affects and emotions correspond to processes of discharge, the final manifestation of which are perceived as feelings. In the present state of our knowledge of affects and feelings we cannot express this difference more clearly.

16. For example, in *The Quest for Consciousness*, Francis Crick and Christoph Koch explained that affect involves 'the more difficult aspects of consciousness' (Crick, 2004, p. xiv); so they 'purposefully ignored [...] how emotions [...] help form and shape the neuronal coalition(s) that are sufficient for conscious perception' and focused instead on 'experimentally more tractable aspects of consciousness'

like visual perception (Koch, 2004, p. 94). Amazingly, they considered the neocortical mechanism of conscious vision to be a simpler problem than the primitive brainstem mechanism of affect, while simultaneously acknowledging that affect shapes the neuronal coalitions that are (supposedly) sufficient for conscious perception. Anyway, even if the problem of affect *is* 'more difficult', that is not a good reason to leave it out of our account of consciousness. Imagine leaving quantum effects out of an account of physics!

17. I do not mean to imply that they neglected it completely; see Hume's (1748) *An Enquiry Concerning Human Understanding*.
18. This term is attributable to Panksepp (2011).
19. Skinner (1953), p. 160.
20. Thorndike (1911).
21. Mazur (2013).
22. See Leng (2018), chapters 16–19, for a fascinating account of the relevant research.
23. See Solms and Panksepp (2010).
24. Panksepp (1974).

Chapter 5: Feelings

1. Alboni and Alboni (2014).
2. The next paragraph paraphrases Merker (2007).
3. See Nummenmaa et al. (2018) – a very interesting article – which supports several of the conclusions I draw below, e.g. regarding the impossibility of 'neutral' affects, the embodiment of affects and the categorical nature of affects.
4. E.g. fear entrains rapid breathing, increased heart rate, redirection of blood from the gut to the skeletal musculature and, thereby, tonic alertness and readiness for escape. If the feeling of fear entrained slow breathing, reduced heart rate, redirection of blood from the skeletal musculature to the gut and, thereby, sluggish repose, *it would not be fear*. The embodied nature of affects is evident from whole-body thermal scans of the different emotions (see Nummenmaa et al., 2018). See also Niedenthal (2007).
5. These philosophical issues are discussed further in Chapter 11.
6. Urologists call this situation 'latch-key urgency'.
7. Ekman et al. (1987).
8. Barrett (2017).

9. See Panksepp (1998) and Panksepp and Biven (2012) for the empirical details and bibliographic references.

10. See LeVay (1993).

11. For those interested in the anatomical details: in typical males the focus of the LUST circuit is the anterior hypothalamus (especially the interstitial nuclei), whence it descends via the bed nucleus of the stria terminalis to the PAG. Chemically, the steroid hormone testosterone (which is released by the testicles and acts largely on the anterior hypothalamus) mediates the release in the brain of a peptide called vasopressin, which accounts for male sexual arousal. In typical females, the ventromedial hypothalamus is the locus of sexual control, and the main chemicals are oestrogen and progesterone (both released by the ovaries, the female equivalent of the testicles). These hormones mediate the activity of oxytocin in the brain, a peptide that governs much of the female-typical sexual response. LUST is mediated also by other peptides such as LH-RH and CCK.

12. SEEKING is mediated also by the neurotransmitter glutamate and a host of peptides, such as oxytocin, neurotensin and orexin.

13. I return to the topic of thinking, in some detail, in Chapter 10.

14. It has its origins in the medial parts of the amygdala and it passes through the bed nucleus of the stria terminalis and the medial and perifornical hypothalamus, on its way to the PAG. Its command neuromodulator is a peptide called substance P, acting together with glutamate and acetylcholine. The latter fact might explain why REM behaviour disorder is so frequently expressed through RAGE.

15. The FEAR circuit arises from the central and basolateral amygdala. Chemically it is mediated by the neurotransmitter glutamate, in addition to the peptides DBI, CRF, CCK, α-MSH and NPY.

16. See Tranel et al. (2006), who explicitly focused on SM's *subjective* emotionality.

17. See Blake et al. (2019).

18. The same thing applies to other similar tasks, such as 'intuition' in the Iowa gambling task (Turnbull et al., 2014). As I explained to Nicholas Humphrey, who objected to my use of the term 'gut feeling' rather than 'guessing' in blindsight, it all depends on what questions you ask (The Science of Consciousness conference, Interlaken, June 2019).

19. LeDoux (1996).

20. Bowlby (1969).

21. E.g. Yovell et al. (2016) and Coenen et al. (2019).

22. Solms and Panksepp (2010).
23. Via the bed nucleus of the stria terminalis, preoptic area and dorsomedial thalamus. In addition to the opioid mechanisms described in the text, PANIC is mediated by the neurotransmitter glutamate and the neuropeptides oxytocin, prolactin and CRF.
24. This is presumably why mental pain is so frequently somatised as physical pain. See Eisenberger (2012) and Tossani (2013).
25. CARE is also mediated by dopamine.
26. The CARE circuit descends from the anterior cingulate, via the bed nucleus of the stria terminalis, preoptic area and ventral tegmental area, to the PAG.
27. Forrester et al. (2018).
28. Panksepp and Burgdorf (2003). See also www.youtube.com/watch?v=j-admRGFVNM.
29. Hopefully, this is no longer a popular game. I am not *endorsing* the fact that native Americans were dominated by settler farmers, but they certainly were.
30. Empathy arises from the intentional stance, or 'theory of mind', which is of course not equally developed in all mammal species. The development of empathy is therefore by no means an automatic process, as the 'mirror neuron' theory might suggest. Empathy is not a reflex; it is a developmental achievement (see Solms, 2017a).
31. However, the dorsomedial thalamus and parafasicular area appear to be especially important, as is the PAG, of course. If anything is the command modulator of PLAY it is mu opioids, but this might simply reflect the fact that safety (i.e. low PANIC/GRIEF) is a necessary precondition for PLAY. Other putative modulators of PLAY are glutamate, acetylcholine and cannabinoids. The thalamocortical circuit identified by Zhou et al. (2017) pertains only to a specific aspect of PLAY, namely dominance. See also Van der Westhuizen and Solms (2015), Van der Westhuizen et al. (2017).
32. Pellis and Pellis (2009).

Chapter 6: The Source
1. Moruzzi and Magoun (1949).
2. Fischer et al. (2016). The lesion was near (just above) the medial parabrachial nucleus. Interestingly, this is the region where Hobson

identified the cholinergic source cells for REM sleep. See also Parvizi and Damasio (2003), Golaszewski (2016).

3. Blomstedt et al. (2008). The electrode was placed in the substantia nigra. The intended site was the subthalamic nucleus.
4. Damasio et al. (2000), Holstege et al. (2003).
5. This refers to the neocortical convexity, not the hippocampus.
6. Garcia-Rill (2017).
7. Holeckova et al. (2006).
8. Use of the terms 'channel' and 'state' in this way is attributable to Mesulam (2000). In Chapter 9 I use the analogy of 'operating modes' for 'state' functions. For a readable account of what the brain actually looks like in different states or operating modes, at the cellular level, see Abbott (2020).
9. Panksepp (1998), p. 314. This quotation from Panksepp continues below and incorporates the following two endnotes.
10. Bailey and Davis (1942).
11. Depaulis and Bendler (1991).
12. See Walker (2017) for a highly readable account of how this works. Even molluscs and echinoderms (like starfish) show a sleep/wake cycle. Panksepp (1998), p. 135, points out that sleep regulation is phylogenetically older than the reticular activating system. On this basis, he makes the intriguing suggestion that:

     What is now the REM sleep mechanism originally mediated the selective arousal of emotionality. Prior to the emergence of complex cognitive strategies, animals may have generated most of their behavior from primary-process psychobehavioral routines that we now recognize as the primitive emotional systems [...] In other words, many of the behaviors of ancient animals may have emerged largely from preprogramed emotional subroutines. These simple-minded behavioral solutions were eventually superseded by more sophisticated cognitive approaches that required not only more neocortex but also new arousal mechanisms to sustain efficient waking functions within those emerging brain areas.

     In light of what I propose below, Panksepp's suggestion can be reworded thus: the forebrain adds to the lower, automatised instinctual motor programmes a capacity to *contextually* modulate emotional behaviour – and thereby to learn from experience.

13. This is not to be equated with the philosophical concept of intentionality or 'aboutness'. Panksepp means something like 'volition'. However, when

I address the philosophical concept later, it will become evident that it is deeply related to volition.

14. Accordingly, in Pfaff's (2005) exhaustive treatment of the topic of arousal, which he describes as 'the most fundamental force in the nervous system', he operationalises the term as follows: '"Generalized arousal" is higher in an animal or human being who is: (S) more alert to sensory stimuli of all sorts, and (M) more motorically active, and (E) more reactive emotionally'. In view of the centrality of 'arousal' in this book, I quote a lengthy extract from Pfaff (2005) in the Appendix on p. 306, which also provides a useful bridge to the topic of the next chapter. I am grateful for this opportunity to acknowledge the seminal work of Donald Pfaff, who (in the early 1990s already, when I first met him) showed an unusual appreciation of Freud's formulation of 'drive'.

15. The PAG projects to all the neuromodulatory source nuclei of the reticular activating system. The other major brainstem destinations of PAG projections are the medial hypothalamus, cuneiform nucleus, pontine reticular formation, solitary nucleus, gracile nucleus, dorsoreticular nucleus and ventrolateral medulla. See Linnman et al. (2012).

16. Venkatraman, Edlow and Immordino-Yang (2017). I do not like the word 'descending' in this context, because the PAG *integrates* higher cerebral and lower visceral affective feedback. It only 'descends' in the sense that it results in motor output. 'Centripetal' would be a better word, which could be contrasted with a 'centrifugal' network (i.e. with what Edlow calls the 'modulatory' network). A 'centripetal' network would include both the 'descending' and 'ascending' networks. Linnman et al. (2012) accordingly describe the PAG as the site of *interaction between* the 'descending limbic' and 'ascending sensory' systems.

17. Venkatraman, Edlow and Immordino-Yang (2017).

18. Ibid.

19. Linnman et al. (2012), p. 517, emphasis added.

20. Ibid., emphasis added. Not surprisingly, in human brain imaging studies, the PAG is found to belong to a 'salience' network (Seeley et al., 2007).

21. Ezra et al. (2015), p. 3468, emphasis added.

22. Panksepp and Biven (2012), p. 413, emphasis added.

23. Linnman et al. (2012), p. 506, emphasis added.

24. Panksepp and Biven (2012).

25. It used to be called the 'central grey'.

26. The 'back' ones are the lateral and dorsolateral PAG. The 'front' one is the ventrolateral PAG. This classification neglects the dorsomedial PAG.

27. See Venkatraman, Edlow and Immordino-Yang (2017):

    When stimulated, this column produces emotional vocalization, confrontation, aggression and sympathetic activation, shown by increased blood pressure, heart rate, and respiration [...] Within this dorsolateral/ lateral column itself, there are two parts. The rostral part is responsible for power/dominance (producing a fight response), while the caudal part invokes fear (producing a flight response) with blood flow to the limbs.

28. The front column

    receives poorly localized 'slow, burning' somatic and visceral pain signals, and on stimulation produces passive coping, long-term sick behaviour, freezing with hyporeactivity and an inhibition of sympathetic outflow [...] In this way, it is likely involved in background emotions such as those that contribute to mood (ibid.).

29. Simone Motta and colleagues put it like this (Motta, Carobrez and Canteras, 2017), p. 39:

    [The PAG] has been commonly recognized as a downstream site in neural networks for the expression of a variety of behaviors and is thought to provide stereotyped responses. However, a growing body of evidence suggests that the PAG may exert more complex modulation of a number of behavioral responses and work as a unique hub supplying primal emotional tone to influence prosencephalic sites mediating complex aversive and appetitive responses.

30. The PAG is a frequent site of deep brain stimulation for the treatment of chronic pain but it does not diminish cortical somatosensory capacities.

31. Merker (2007). He, in turn, attributes this insight to Penfield and Jasper (1954).

32. I am simplifying the technical term 'mesodiencephalic selection triangle' for the purposes of this book. Also, my usage of Merker's term refers to a decision *interface* more than a *triangle* (i.e. an interface between need and context). As we shall see in the next chapter, action ('action selection' for Merker) and perception ('target selection' for Merker) – which together make up the context – are two sides of the same coin. Here is Merker's (2007), p. 70, description of it:

    However much the telencephalon subsequently expanded, even to the point of burying the mesodiencephalon under a mushrooming mammalian neocortex, no other arrangement was ever needed, and that for the most fundamental of reasons. No efferent nerve has its motor nucleus situated above the level of the midbrain. This means that the very narrow cross-section of the brainstem at the junction between midbrain and diencephalon [...] carries the total extent of information by which the forebrain is ever able to generate, control, or influence behaviour of any kind.

Merker calls this cross-section the 'synencephalic bottleneck'. He adds:

One need not know anything more about the vertebrate brain than the fact that its most rostral motoneurons are located *below* the synencephalic bottleneck, to know that the total informational content of the forebrain must undergo massive data reduction in the course of its real-time translation into behaviour.

The fact that the decision as to what to do next is made at this (brainstem) level – i.e. after the forebrain regions have submitted their 'bids' – is dramatically illustrated by the example of suffocation alarm, discussed above: all cognitive considerations are *overridden* by the feeling of air hunger, which is triggered at the level of the brainstem. The respiratory control centres are located in the pons and medulla oblongata.

33. Panksepp (1998), p. 312, described the functional arrangement of the SELF, a decade before Merker, as follows:

The deeper layers of the colliculi constitute a basic motor mapping of the [objective] body, which interacts not only with visual, auditory, vestibular and somatosensory systems but also with nearby emotional circuits of the PAG. The PAG elaborates a different, visceral type map of the [subjective] body along with basic neural representations of pain, fear, anger, separation distress, sexual and maternal behaviour systems (as summarized [in the previous chapter of] this book). Adjacent to the PAG is the mesencephalic locomotor region, which is capable of instigating neural patterns that would have to be an essential substrate for setting up various coherent action tendencies.

Damasio and Carvalho (2013) also proposed this general functional arrangement for what they called the 'proto-self'. They pointed out that other primitive heteromodal sensory maps of the body are provided by the parabrachial complex and nucleus of the solitary tract. These might plausibly be evolutionary precursors of the integrative function performed by the midbrain decision triangle.

34. This from Merker (2007), p. 73:

[The conscious self] is single, and located behind the bridge of the nose inside your head. From there we *appear* to confront the visible world directly through an empty and single cyclopean aperture in the front of our head (Hering 1879; Julesz 1971). Yet that is obviously a mere appearance, since if we were literally and actually located inside our heads we ought to see, not the world, but the anatomical tissues inside the front of our skulls when looking. The cyclopean aperture is a convenient neural fiction through which the distal visual world is 'inserted' through a missing part of the proximal visual body, which is 'without head' as it were or, more precisely, missing its upper face region (see Harding 1961). Somesthesis by contrast maintains unbroken continuity across this region. The empty opening through which we gaze out at the world betrays the simulated nature of the body and world that are given to us in consciousness.

35. Ibid., p. 72. See Stoerig and Barth (2001) for a plausible simulation. This gives some impression of the likely sensory-motor world of the hydranencephalic child and decorticate animal (cf. also blindsight).

36. White et al. (2017), Panksepp (1998), p. 311.

37. Merker (2007), p. 72.

38. Hohwy (2013).

39. As he actually did, within a few years of his operation, from an (otherwise easily treatable) upper respiratory infection that was recognised too late.

40. See Friston (2005).

41. Hohwy (2013), Clark (2015).

42. Consider, for example, the dorsal paired medial neurons in the fruit fly *Drosophila*. Surprisingly, there even seem to be primitive precursors in the nematode *Caenorhabditis elegans* (see Bentley et al., 2016; Chew et al., 2018).

43. The following image from Freud (1925), p. 231, might help in picturing the functional arrangement just described, and may simultaneously enable us to replace his 'metapsychological' terms with physiological ones:

> Cathectic innervations are sent out and withdrawn in rapid periodic impulses from within into the completely pervious system Pcpt-Cs [the 'perceptual consciousness' system]. So long as that system is cathected in this manner it receives perceptions (which are accompanied by consciousness) and passes the excitation onwards to the unconscious mnemic systems; but as soon as the cathexis is withdrawn, consciousness is extinguished and the functioning of the system comes to a standstill.

> It is as though the [id] stretches out feelers, through the medium of the system Pcpt-Cs, towards the external world and hastily withdraws them as soon as they have sampled the excitations coming from it.

'Cathexis' is modulatory arousal. Therefore, the 'cathectic innervations' palpating cortical perception in this figurative image are pulses of core brain arousal. I have replaced Freud's term 'the unconscious' with 'the id' in this quotation, to circumvent the fact that he wrongly conflated these two systems (Solms, 2013).

44. Merker (2007). Panksepp and Biven (2012), pp. 404–5, use the hippocampal theta rhythm as an example of how this 'common currency' might pan out physiologically, bearing in mind that the hippocampus encodes *context*:

> There are suggestive hints in the traditional neuroscience literature for certain types of relevant synchronous oscillation within the brain, such as the 4–7 Hz rhythms in the hippocampus known as *theta rhythm*, which helps animals to investigate the world (e.g. sniffing in rats) and thereby create memories in the hippocampus. The theta rhythm is the highly characteristic neural signature of the hippocampus as it is actively processing information. This rhythm is especially evident during artificial arousal of the SEEKING system in rats, a premier information-gathering emotional system, as animals sniff and investigate their surroundings (Vertes and Kocsis, 1997). In other words, the sniffing rhythm typically corresponds to the ongoing frequency of the hippocampal theta [...] This may highlight how cognitive knowledge emerges from the patterned arousals of the affective processes.

45. 'Touch' is the colloquial term for somatic sensation, which contains submodalities like muscle and joint sense, temperature sense, vibration sense – as do other perceptual modalities.

46. When they are salient.

47. Solms (2013).

48. Acetylcholine may be said to modulate confidence in error signals, but this is a huge simplification. Equivalently broad generalisations could be made about serotonin in relation to prediction signals, dopamine in relation to active states, and noradrenaline in relation to sensory states. See Parr and Friston (2018) for a more elaborated view.

49. 'Physics' (φυσική) means 'knowledge of nature' – all of nature – not knowledge only of things you can see and touch. It provides the most fundamental explanation of natural phenomena. Many people assume that physics studies only matter – and therefore excludes the mind by

definition – but this would imply that the mind is not part of nature, which begs the very question this book is about.

'A natural phenomenon is fully explained physically only when one has traced it back to the ultimate forces of nature that ground it and are effective in it' (Helmholtz, 1892). Matter turns out to be an energy state (hence $E=MC^2$). 'Ultimate forces' *explain* superficial phenomena; they are not observed directly, they are inferred. For this reason, they are reported scientifically in non-phenomenal terms, as abstractions.

Cf., Kant's *Critique of Pure Reason*:

Experience itself – in other words, empirical knowledge of appearances – is thus possible only in so far as we subject the succession of appearances, and therefore all alteration, to the law of causality; and as likewise follows, the appearances, as objects of experience, are themselves possible only in conformity with the law.

Mathematical abstractions are conventionally preferable to verbal ones, because they require the inferred forces (and the relations between them) to be measurable and quantifiable. This provides a common numerical currency beneath the variegated phenomenal surfaces whereby the lawful relations between the 'ultimate forces' of nature can be calculated. As Galileo said: 'The book of nature is written in the language of mathematics.'

Chapter 7: The Free Energy Principle

1. Crystals minimise their free energy in a trivial way because their nonequilibrium steady state has a point attractor. That is, they just arrange themselves into compact patterns and stay there, even when slightly disturbed. Things get more complex when the attracting set has an itinerant structure with dynamics of the sort that the human brain conforms to.

2. In brief, your h-index is the number of your publications that have been cited (by your peers) more times than their ranking in the sequence of your cited publications. Thus, if your fortieth most-cited publication has been cited forty-two times, but your forty-first most-cited publication has been cited only thirty-nine times (i.e. less than forty-one), then your h-index is forty.

3. As of 20 July 2020.

4. Carhart-Harris and Friston (2010).

5. Freud (1894), p. 60.

6. Friston (2005), emphasis added. The same applied even to his article with

Carhart-Harris, which begins like this (Carhart-Harris and Friston, 2010, p. 1265, emphasis added):

Freud's descriptions of the primary and secondary processes are consistent with self-organized activity in hierarchical *cortical* systems and [...] his descriptions of the ego are consistent with the functions of the default-mode and its reciprocal exchanges with subordinate brain systems. This neurobiological account rests on a view of the brain as a hierarchical inference or Helmholtz machine. In this view, large-scale intrinsic networks occupy supraordinate levels of hierarchical brain systems that try to optimize their representation of the sensorium. This optimization has been formulated as minimizing a free-energy; a process that is formally similar to the treatment of energy in Freudian formulations.

7. Friston (2013).

8. Solms and Friston (2018).

9. Parvizi and Damasio (2003).

10. See Damasio (2018) and my review of it, Solms (2018a).

11. The insight of Einstein's that I am referring to is the quantum nature of light (Einstein, 1905). Stephen Hawking replies:

Not only does God definitely play dice, but He sometimes confuses us by throwing them where they can't be seen. Many scientists are like Einstein, in that they have a deep emotional attachment to determinism. Unlike Einstein, they have accepted the reduction in our ability to predict, that quantum theory brought about.

It turns out that the phenomena of consciousness, too, are predictable only *probabilistically*.

12. This was at USC in April 2018. In a phone call in January 2019, after he read the published version of our article, he was still defending 'our science' on the basis that consciousness was intrinsically *biological*. Happily, however, he has changed his mind since then; see Man and Damasio (2019).

13. Interestingly, Hermann von Helmholtz (one of Johannes Müller's pupils who founded the Berlin Physical Society) played a big part in the formulation of this law.

14. This is true in reality, within the thermodynamic limit, but not (strictly speaking) in theory. It would be more correct theoretically to say: The Second Law stipulates that natural processes are very, very *unlikely* to be reversible. This is because, when you solve the equations of any system, you don't only need the relevant dynamic laws, you also need the *initial conditions* of the system; and these break the symmetry of the

fundamental equations. Therefore, if you could start with all the pieces of a shattered cup moving towards each other at exactly the right speed, and with the sound waves moving back towards the cup in exactly the right way, and likewise with the energy that had moved through the floor when it landed, then everything could come back together to recreate the initial unbroken cup. However, because we are never in that situation, we don't ever see entropy decreasing in reality.

15. Except for very, very brief periods of time.

16. When heat is added to a substance, the molecules and atoms vibrate faster. As atoms vibrate faster, the space between them increases. The motion and spacing of the particles determine the state of matter in the substance. The end result of increased molecular motion is that the substance expands and takes up more space.

17. Technically, entropy is associated with wasted energy, but it is not the same thing. Entropy is dimensionless, whereas energy has dimensions.

18. Technically, it is associated with the number of equivalent states corresponding to a given macroscopic configuration.

19. The physicist Alan Lightman (2018), pp. 67–8, puts it beautifully:

    If we relentlessly divide space into smaller and smaller pieces, as did Zeno, searching for the smallest element of reality, once we arrive at the phantasmagoric world of Planck, space no longer has meaning. At least, what we *understand* as 'space' no longer has meaning. Instead of answering the question of what is the smallest unit of matter, we have invalidated the words used to ask the question. Perhaps that is the way of all ultimate reality, if such a thing exists. As we get closer, we lose the vocabulary.

    Another physicist, Carlo Rovelli (2014), p. 167, provides a more prosaic account:

    The backdrop of space has disappeared, time has disappeared, classic particles have disappeared, along with the classic fields. So what is the world made of? The answer now is simple: […] the world is made entirely from quantum fields.

    Even quantum uncertainty is part of the *physical* universe.

20. A 'byte' comprises eight bits. A 'gigabyte' therefore has 8 billion bits. The *speed* of information processing is expressed in 'gigahertz', one GHz being 1 billion bit-flips per second. The physical reality of information is reflected by the fact that these units are both measurable and purchasable (from internet service providers, for example).

21. See the Appendix on p. 306. Here one must define who is uncertain of

what. The same applies to 'communicates' in the next-but-one sentence: who is communicating with whom? Please read on.

22. Gosseries et al. (2011).

23. If you think this conflates information entropy with thermodynamic entropy, see the quote from Tozzi, Zare and Benasich in note 61below.

24. As you will see, the fact that we have *multiple* needs, including SEEKING, is crucial.

25. Consider, for example, quantum entanglement between two particles: the one particle 'carries information' about the other.

26. Jaynes (1957). Increasing the size of the system described in our formal example above (the gas in the chamber) increases its thermodynamic entropy because it increases the number of possible microstates of the system that are consistent with the measurable values of its macroscopic variables, thus making any complete description of its state more information-heavy. To be exact: in the discrete case using base two logarithms, the reduced thermodynamic entropy is equal to the minimum number of yes/no questions needed to be answered in order to fully specify the microstate, given that we know the macrostate. A direct and physically real relationship between thermodynamic entropy and information entropy can be found by assigning a unit of measurement to each microstate that occurs per unit of a homogeneous substance, then calculating the thermodynamic entropy of these units. By theory or by observation, the microstates will occur with different probabilities and this will determine the information entropy. This shows that Shannon's entropy is a true statistical measure of microstates that *does not have a fundamental physical unit other than the units of information.*

27. Shannon (1948).

28. For example, Chalmers's 'Double-aspect principle', discussed in Chapter 11.

29. Wheeler was a student of Niels Bohr, who formulated the principle of complementarity. The principle holds that objects have complementary properties which cannot be *observed* or *measured* simultaneously. Examples of complementary properties are particle and wave, position and momentum, energy and duration, spin on different axes, value of a field and its change (at a certain position), and entanglement and coherence.

30. His actual phrase was 'it from bit'. Here is the original quotation (Wheeler, 1990, pp. 310–1):

It from Bit. Otherwise put, every it – every particle, every field of force, even the space-time continuum itself – derives its function, its meaning, its very existence entirely – even if in some contexts indirectly – from the apparatus-elicited answers to yes or no questions, binary choices, bits. *It from Bit* symbolizes the idea that every item of the physical world has at bottom – at a very deep bottom, in most instances – an immaterial source and explanation.

31. Ibid. The words quoted here follow immediately from the quotation above.

32. Attentive readers might notice here that I am finally addressing the question I asked myself in childhood, about the first sunrise, as discussed in the Introduction.

33. Darwin (1859). See Friston (2013). Rovelli (2014), pp. 225–6, provides a lucid explanation of this:

   A living organism is a system that continually re-forms itself, interacting ceaselessly with the external world. Of such organisms, only those continue to exist which are more efficient at doing so and, therefore, living organisms manifest properties which have suited them for survival. For this reason, they are interpretable, and we interpret them, in terms of intentionality, of [aim and] purpose. The finalistic aspects of the biological world (this is Darwin's momentous discovery) are therefore the result of the selection of complex forms effective in persisting. But the effective way of continuing to exist in a changing environment is to manage correlations with the external world better, that is to say, information; to collect, store, transmit and elaborate information. For this reason, DNA exists, together with immune systems, sense organs, nervous systems, complex brains, languages, books, the library of Alexandria, computers and Wikipedia: they maximise the efficiency of information management – the management of correlations favouring survival.

34. What matters here is not that self-organising systems are always alive (they are not) but rather that living systems are always self-organising. Even more importantly: not all self-organising systems are conscious. Regarding the physics of the origins of life, see England (2013) for an interesting treatment along similar lines to that of Friston. England's 'dissipation-driven adaptation' theory argues that groups of atoms that are driven by external energy sources tend to tap into those sources, aligning and rearranging, so as to better absorb the energy and dissipate it as heat (i.e. they minimise their own entropy at the expense of their environment). He further shows that this dissipative tendency fosters self-replication: 'A great way of dissipating more is to make more copies

of yourself.' Regarding the *biology* of the origin of life, though, see Lane (2015).

35. Ashby (1947). See also Conant and Ashby (1970).

36. Friston (2013), p. 6. What the computer monitor displayed were, of course, *representations* of the subsystems, which should not be confused with the statistical dynamics themselves. The same applies to the perceptual qualia associated with cortical functioning, as we shall see.

37. Andrey Markov (1856–1922) was a brilliant Russian mathematician, as were his brother and his son. He worked mainly on stochastic processes and is now famous for what became known as Markov chains and Markov processes. He was a consummate rebel. The institutional pushback was such that he never received academic recognition during his lifetime. In 1912 he protested Leo Tolstoy's excommunication from the Russian Orthodox Church by requesting his own excommunication. The Church complied with the request.

38. By 'vice-versa' I do not mean that the system's internal states are hidden from external (not-system) states, I mean that the system's *point of view* is hidden; it is available only to itself. When it comes to the problem of other minds, one system can never know the internal states of another system not only because all external states are hidden from it but also because the internal states of other systems are internal to those systems alone.

39. Friston (2013), p. 8.

40. The term 'autopoiesis' was introduced by Maturana and Varela (1972) to define the self-maintaining chemistry of living cells.

41. Considering what I said above about the representational capacity of the system, philosophical readers will observe that it also brings about 'intentionality' in Brentano's (1874) sense.

42. Friston (2013), p. 2.

43. Or simply hidden and not available for effective work.

44. That is what Helmholtz called the energy that isn't free (*TS* in the equation). The distinction between 'free' and 'bound' energy will surely not be lost on Freud scholars. See Chapter 10, note 16, re Freud's notion of 'secondary process'.

45. $A = U + pV - TS$; where p denotes pressure and V denotes volume. This equation quantifies the free energy of systems whose work is associated with system expansion or compression at constant temperature and pressure.

46. The equation comes in a long and a short form. Here is the long one: $F(s, \mu) = Eq[-\log p(s, \psi \mid m)] - H[q(\psi \mid \mu)]$, where F ('variational' free energy, or Friston free energy for short) is the isomorph of Gibbs free energy and Helmholtz free energy. Regarding the other quantities in the equation, s denotes sensory states (of the Markov blanket, discussed above), μ denotes internal states, Eq denotes average energy, $p(s, \psi \mid m)$ denotes a probability density over sensory and external (hidden) states under a generative model m, H denotes entropy, and $q(\psi \mid \mu)$ denotes a variational density over the hidden states parametrised by internal states. The relationship between this equation used in information science and the one used in thermodynamics is not immediately obvious. When the long equation is compressed like this, however, it looks more like the thermodynamic one: $F = Eq - H$. Here, F denotes Friston free energy, Eq denotes average energy and H denotes entropy.

47. I.e. the average information gained on many measurements of microstates.

48. The further expressions quoted above (concerning surprisal and divergence) basically tell us that, whereas Helmholtz free energy is a measure of the energy available to do effective work, Friston free energy is a measure of *the difference between the way* the *world is modelled by a system and the way the world actually behaves*. (I will explain how this relates to work in a moment.)

49. The complexity term of Friston free energy shares the same fixed point as Helmholtz free energy (under the assumption that the system is thermodynamically closed but not isolated). If sensory perturbations are suspended for a suitably long period of time, complexity is minimised because accuracy can be neglected. At this point, the system is at equilibrium and internal states minimise Helmholtz free energy by the principle of minimum energy (which is basically a restatement of the Second Law).

50. Formally stated: surprisal is the negative log of how probable the outcome (s) is, under a given model of the hidden states of the world ($[-\log p(s, \psi \mid m)]$ in the equation above).

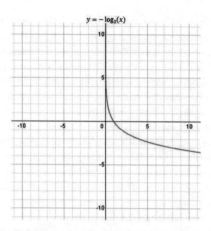

$y = -\log_2(x)$

As the probability (x axis) heads towards 0, so the surprisal (y axis) goes up; as the probability heads towards 10, so the surprisal goes down.

51. Thus, minimising surprisal about things that have happened will, on average, minimise entropy. (Surprisal is an attribute of data or observations, while free energy is an attribute of beliefs. Therefore, the entropy part of free energy is not the average surprisal of the observations, it is the entropy of beliefs about the latent causes of observations.)

52. Friston free energy (average energy minus entropy) is equivalent to surprisal (which is expressed as $-\log p(s \mid m)$) plus 'perceptual divergence' (which is expressed as $DKL\ [q(\psi \mid \mu) \parallel p(\psi \mid s, m)]$), which is always greater than or equal to surprisal alone. Average 'surprisal' essentially means (information) entropy, as I explain below. 'Perceptual divergence' measures the difference between hypothesised events and actual events, under a generative model. DKL denotes perceptual divergence. It stands for Kullback–Leibler divergence, also known as relative entropy, which quantifies the divergence between two probability densities: the two that interest us are about hidden states – the variational density encoded by internal (e.g. neuronal) states and the actual conditional density, given sensory states. DKL is always greater than or equal to zero. Intuitively, this is because negative log functions (explained below) always have roughly U-shaped plots, so a line joining two points on the U can never be lower than the bottom of the U (technically, this is called a 'concave up function'). This ensures that free energy places an upper bound on surprisal.

53. See Clark (2017):

    Self-evidencing [...] occurs when a hypothesis best explains some piece of evidence and, in virtue of that explanatory success, thereby provides evidence for its own truth or correctness. In such cases, the occurrence of the evidence is best explained by the hypothesis but the fact that the evidence occurs at all is used to lend support to the hypothesis itself.

54. There is another technical twist here that discloses the deep relationship between self-organisation, information theory and existential inference. Surprisal is just negative log likelihood of some sensory states given a model within a Markov blanket. In Bayesian statistics, that is also known as (the logarithm of) model evidence. This is what licenses the minimisation of Friston free energy to be described as *self-evidencing*; i.e. maximisation of evidence for a model that generates the sensory states of any system that features a Markov blanket.

55. It might therefore be more correct to say: 'the answer to this question must determine the average behaviour of the system'.

56. Please note: this mechanism places probabilistic constraints on 'free will'. You are free to enter the lion's den, for example, but you are *unlikely* to do so, and if you do, you will *probably* die. See Bayes's theorem, below.

57. Where A and B are events, $P(A \mid B)$ is the likelihood of A occurring given that B is true (the conditional probability), $P(B \mid A)$ is the likelihood of B occurring given that A is true (another conditional probability), and $P(A)$ and $P(B)$ are the probabilities of observing A and B independently of each other (the marginal probability). My verbal translation of the theorem follows Joyce (2008), who wrote $P(A) \mid P(B)$ as a separate ratio.

58. Bayes's theorem can be reformulated in free-energy terms – whereby free energy is decomposed into 'accuracy' and 'complexity'. Model evidence is the difference between accuracy and complexity, since models with minimum free energy provide accurate explanations of data under complexity costs, which in turn means that reducing model complexity improves model generalisability but at the cost of accuracy. (In Bayesian terms, 'likelihood' must be assessed in relation to 'probability' to prevent overfitting.) Knill and Pouget (2004), p. 713, get to the heart of the matter: 'The real test of the Bayesian coding hypothesis is in whether the neural computations that result in perceptual judgements or motor behaviour take into account the uncertainty available at each stage of processing.'

59. For a detailed account of the functional micro-anatomy of this process, see Friston (2005), Adams, Shipp and Friston (2013) and Parr and Friston (2018).

60. Thus, for example, prediction errors appear to be communicated using gamma-range (high) frequencies while predictions seem to be conveyed by beta-range (lower) frequencies. See Bastos et al. (2012, 2015).

61. See Tozzi, Zare and Benasich (2016):

> Minimizing variational free-energy necessarily entails a metabolically efficient encoding that is consistent with the principles of minimum redundancy and maximum information transfer (Picard and Friston, 2014). Maximizing mutual information and minimizing metabolic costs are two sides of the same coin; by decomposing variational free energy into accuracy and complexity, one can derive the principle of maximum mutual information as a special case of maximizing accuracy, while minimizing complexity translates into minimizing metabolic costs (Friston et al., 2015). Thus, the basic form of Friston's free-energy principle supports the idea that the energetic levels of spontaneous brain activity, which are lower when compared with evoked activity, allow the CNS to obtain two apparent contradictory achievements: to minimize as much as possible the metabolic costs, and to the largest extent possible, maximize mutual information.

62. See Clark (2015), p. 268. However, this must be balanced against what he says about the default activity of the SEEKING system (p. 263):

> Such creatures are built to seek mates, to avoid hunger and thirst, and to engage (even when not hungry and thirsty) in the kinds of sporadic environmental exploration that will help prepare them for unexpected environmental shifts, resource scarcities, new competitors, and so on. On a moment-by-moment basis, then, prediction error is minimized only against the backdrop of this complex set of creature-defining 'expectations'.

63. Friston and Stephan (2007), p. 427.

64. I based this fable on Clark's (2015) metaphor of a leaky dam.

Chapter 8: A Predictive Hierarchy

1. Hohwy (2013), p. 63.

2. I say 'core' rather than 'top' of the hierarchy because it seems odd (anatomically) to place hypothalamic and brainstem body-monitoring nuclei *above* the neocortex. I prefer to picture the hierarchy as unfolding concentrically, from inside to outside, something like the layers of an onion (see Mesulam, 2000). After all, as every embryologist knows, the nervous system is a tube. It should be remembered also that the neural

tube is formed from the ectoderm, through invagination of the neural plate, whereby the central canal takes the place of the world outside.

3. This is not the standard predictive-coding view, but that is the whole point of this book: I am showing how our understanding of the brain has been hampered by cortico-centrism. Cf. Pezzulo (2014), p. 910: 'An interesting direction for future research would be to look at homeostatic regulation within the active inference framework.'

4. Clark (2015), p. 21, provides a succinct summary, using vision as his model example:

> A certain pattern of retinal stimulation, encountered in a given context, might be best accounted for using a generative model that [...] combines top-level representations of interacting agents, objects, motives, and motions with multiple intermediate layers capturing the way colours, shapes, textures, and edges combine and temporally evolve. When the combination of such hidden causes (which span many spatial and temporal scales) settles into a coherent whole, the system has self-generated the sensory data using stored knowledge and perceives a meaningful, structured scene. It is again worth stressing that this grip upon the structured distal scene must be generated using only the information available from the animal's perspective. It must be a grip, that is to say, rooted entirely in the combination of whatever pre-structuring (of brain and body) may be present thanks to the animal's evolutionary history and the plays of energetic stimulation that have been registered by the sensory receptors. A systematic means of achieving such a grip is provided by the ongoing attempt to self-generate the sensory signal using a multilevel architecture. In practice, this means that top-down and lateral connections within a multilevel system come to encode a probabilistic model of interacting causes operating at multiple scales of space and time. We recognize objects and states and affairs [...] by finding the most likely set of interacting factors (distal causes) whose combination would generate (hence predicts, and best accounts for) the incoming sensory data.

5. Cf. Gregory (1980).

6. This paragraph paraphrases Hohwy (2013), pp. 81–2, but corrects his exteroceptive bias.

7. I do not mean that homeostatic settling points *never* change, only that their degrees of freedom are extremely limited; therefore, almost all allostasis involves behavioural change rather than the updating of innate priors.

8. I am skipping over the matter of 'empirical Bayes' here. It is possible to learn from scratch, but it is terribly expensive, biologically.

9. To illustrate the extent of the problem, consider Hohwy (2013), p. 206:

> In spite of telling the full prediction error story, there seems however to be no contradiction involved in conceiving a creature with all the machinery for prediction error minimization who engages in the same perceptual inference as we do – to whatever degree of natural, neuronal detail we care to specify – and yet who is not phenomenally conscious. We would expect the creature to be conscious, of course, but nothing in the total physical story entails that it will be. This means it is left an open question whether consciousness is something over and above the physical or not.

How could a creature with exactly the same neural machinery as you and me *not* be conscious? I submit that the 'philosophical zombie' Hohwy imagines here lacks the machinery for feelings. For an alternative take on the problem he raises, see Chapter 12.

Chapter 9: Why and How Consciousness Arises

1. Rolls (2019), p. 10, emphasis added. See also my commentary: Solms (2019b).

2. 'More things should not be used than are necessary.' If there are several possible ways that something might be explained, the one using the fewest guesses is probably correct.

3. Roger Sessions, *New York Times*, 8 January 1950.

4. Friston (2009), p. 299, defines the role of precision as being to 'control the relative influence of prior expectations at different levels'. As Hohwy (2013), p. 199, points out:

> If the gain [precision] on one signal is turned up, then the gain on other signals must be turned down. Otherwise the notion of gain is meaningless: weights must sum to one. So, as expectations for precision turn up the gain on one prediction error, the gain for others will turn down.

Cf. Clark (2016), p. 313:

> Feldman and Friston (2010) point out that precision behaves as if it were itself a limited resource, in that turning up the precision on some prediction error units requires reducing it on others. They also comment, intriguingly (op cit., p. 11): 'The reason that precision behaves like a resource is that the generative model contains prior beliefs that log-precision is redistributed over sensory channels in a context-sensitive fashion but is conserved over all channels.'

This evokes what I speculated in Chapter 6 about each neuromodulator adjusting different aspects of arousal. Interestingly, gamma (error) oscillations respond to acetylcholine.

5. Neurophysiologists also call precision 'gain'. The plethora of terminologies used by scientists from different disciplines can be confusing here. Synaptic 'activity' (= neurotransmission) is modulated by post-synaptic 'gain' (= neuromodulation) which, over time, determines 'efficiency' (= neuroplasticity). These same variables are also described by computational scientists as signal 'states', signal 'precisions' and signal 'parameters', respectively. Broadly speaking, neurotransmission entails states, neuromodulation entails precisions and plasticity entails parameters.

6. Friston calls this 'epistemic foraging'.

7. This paragraph closely paraphrases Clark (2015), p. 70.

8. This gives rise to 'inattentional blindness', the best-known example of which is the following: https://www.youtube.com/watch?v=vJG698U2Mvo (see Chabris and Simons, 2010).

9. Varela, Thompson and Rosch (1991), p. 198. Varela calls this approach to perception 'enactive'. See Clark (2015), p. 173:

> The overall concern of an enactive approach to perception is not to determine how some perceiver-independent world is to be recovered; it is, rather, to determine the common principles or lawful linkages between sensory and motor systems that explain how action can be perceptually guided in a perceiver-dependent world.

10. There is a large body of experimental literature that empirically supports this formal conceptualisation of precision. I will mention just one study, for now, which illustrates the basic point. Using fMRI, Hesselmann et al. (2010) looked at base-level activity in the brain under conditions that manipulated precision expectations. They presented (visual and auditory) stimuli with varying levels of noise to research participants; when the participants expected precision, gain was turned up, when they expected imprecision the gain was turned down. That is, top-down prediction was facilitated when imprecise sensory signals were expected. See Feldman and Friston (2010) for a full account of how precision optimisation explains both endogenous and exogenous attention.

11. As you may recall from p. 34, Freud wrote the following at the time in a letter to Fliess of 20 October 1895:

> In the course of a busy night [...] the barriers were suddenly raised, the veils fell away, and it was possible to see through from the details of the

neuroses to the determinants of consciousness. Everything seemed to fit together, the gears were in mesh, the thing gave one the impression that it was really a machine and would soon run of itself.

He went on:

The three systems of neurons [φ, ψ and ω], the free and bound conditions of quantity [$Q\eta$], the primary and secondary processes, the main trend and the compromise trend of the nervous system, the two biological rules of attention and defence, the indications of quality [ω], reality and thought, the state of the psychosexual groups, the sexual determination of repression, and, finally, the determinants of consciousness as a perceptual function – all this fitted together and still fits together! Of course, I cannot contain myself with delight.

Soon after, however, he realised that the gears were not in mesh after all, the thing had not completely fitted together. He then abandoned the project. When Friston and I resurrected it – and, building on more than a century of progress in neuroscience, hopefully completed it (at least in rough outline; see Solms, 2020) – I remembered Freud's poignant remark: 'the thing gave one the impression that it was really a machine and would soon run of itself'. See Chapter 12.

12. Note that ω is not a vector but a scalar (it scales a precision matrix). Note also that the terms Q and $Q\eta$ were used slightly differently in Solms and Friston (2018) from how I am using them here.

13. Where e is a vector and F is a scalar. The dot notation in the equations below implies a dot product (i.e. matrix or vector multiplication). These equations are entirely attributable to Friston's part in our collaboration.

14. $F \approx -\log P(\varphi(M))$.

15. $E[F] \approx E[-\log P(\varphi)] = H[P(\varphi)] = -\frac{1}{2} \cdot log(|\omega|)$, where $E[\cdot]$ denotes expectation or averaging, and P and H denote probability and entropy, respectively (as they did before); under Gaussian assumptions about random fluctuation.

16. The general importance of precision optimisation was recognised long ago, especially in relation to attention (see Friston, 2009; Feldman and Friston, 2010). However, the first scientist to recognise its importance for consciousness itself was Katerina Fotopoulou (2013), p. 35:

One core aspect of consciousness may serve to register the aforementioned quality of 'uncertainty' and its inverse quality, precision. This view goes against the intuitive, long-standing view of core affective consciousness as monitoring hedonic quality, expressed by Solms in Freudian terms as the pleasure–unpleasure series. Instead, I propose that the core quality of this aspect of consciousness (as opposed to perceptual consciousness [...]) is a kind of certainty–uncertainty, or disambiguation principle.

For the record, regarding pleasure–unpleasure vs certainty–uncertainty, I think Fotopoulou slightly misunderstood me, hence my reply: 'I am [...] not sure what she means when she says that affect monitors uncertainty rather than hedonic quality. To my way of thinking, hedonic quality *is* our measure of uncertainty' (Solms, 2013, p. 81).

17.
$$\frac{\partial}{\partial t}M = -\frac{\partial F}{\partial M} = -\frac{\partial F}{\partial e}\frac{\partial e}{\partial M} = \frac{\partial \varphi}{\partial M} \cdot \omega \cdot e \qquad (1)$$

$$\frac{\partial}{\partial t}Q\eta = -\frac{\partial F}{\partial Q\eta} = -\frac{\partial F}{\partial e}\frac{\partial e}{\partial Q\eta} = -\frac{\partial \psi}{\partial Q\eta} \cdot \omega \cdot e \ (2)$$

$$\frac{\partial}{\partial t}\omega = -\frac{\partial F}{\partial \omega} = \tfrac{1}{2} \cdot (\omega^{-1} - e \cdot e) \qquad (3)$$

Where ∂ denotes a partial derivative and t denotes time, and prediction error and free energy are:

$$e = \varphi(M) - \psi(Q\eta)$$
$$F = \tfrac{1}{2} \cdot (e \cdot \omega \cdot e - \log(\omega))$$

I am presenting these equations in broad brushstrokes, as high-level cartoons. They need further development, some of which can only be done when they are implemented (see Chapter 12). For example, as stated above, in fuller treatments, one would also consider hierarchical generative models (with precisions at each level) and accommodate conditional uncertainty about external states. Furthermore, the equations lump all sensory prediction errors together – including exteroceptive, proprioceptive and interoceptive modalities. Note that the term 'proprioceptive' here is synonymous with 'kinaesthetic' (Friston uses 'proprioceptive' simply for alliterative harmony with 'exteroceptive' and 'interoceptive').

18. Technically, this is called a 'gradient descent', where the gradient is the rate of change of free energy with precision.

19. Under our simplifying assumptions about the encoding of Bayesian beliefs (see above).

20. In this sense, precision may be said to play the role of 'Maxwell's daemon' – a thought experiment created by the physicist James Clerk Maxwell (1872): A daemon controls a small door between two

chambers of gas. The gas molecules float about at different speeds. As the faster ones in the first chamber reach the door, the daemon opens and shuts it, very briefly, so that they pass to the second chamber, while the slower ones remain behind. Since faster molecules generate more heat than slower ones, this will decrease the entropy – something that cannot happen without work. If we equate the passage of molecules in this analogy with the neurotransmission of sensory signals, then precision-weighting (neuromodulation) does what Maxwell's daemon does: it selects sensory signals to confound the Second Law. Note that, in terms of the system dynamics described here, consciousness is *the activity of Maxwell's daemon itself*; it is not *the passage of molecules* that is enabled by it. This means that consciousness is *the optimisation of precision with respect to free energy*; it is not *the passing of messages through a predictive hierarchy*. In Figure 17, therefore, consciousness is the activity of ω (precision), which determines the relative influence of e (error signals) over $Q\eta$ (the internal model). The precision-daemon modulates the influence of errors in relation to the model. Consciousness rests on an attribute of beliefs as opposed to the content of beliefs (i.e. it rests on the fluctuating precision of [or confidence in] beliefs about internal and external states of affairs). The activity of this daemon *causes* sensory sequelae through the amplification or attenuation of prediction errors; precision optimisation does not inhere in the error signals themselves. For more on the biological implications of Maxwell's daemon, see the excellent book by Paul Davies (2019) which, unfortunately, appeared after our own application of the concept to consciousness (Solms and Friston, 2018), so we did not cite it.

21. Brown et al. (2013), Feldman and Friston (2010), Frith, Blakemore and Wolpert (2000).

22. Cisek and Kalaska (2010), Frank (2005), Friston et al. (2012), Friston, Schwartenbeck, FitzGerald et al. (2014), Moustafa, Sherman and Frank (2008).

23. Hohwy (2013), Seth (2013), Ainley et al. (2016). In relation to interoceptive sensitivity and the social modulation of pain, see also Crucianelli et al. (2017), Fotopoulou and Tsakiris (2017), Krahé et al. (2013), Decety and Fotopoulou (2015), Paloyelis et al. (2016) and Von Mohr and Fotopoulou (2017).

24. Ferrarelli and Tononi (2011), Lisman and Buzsaki (2008), Uhlhaas and Singer (2010).

25. See Hobson (2009) and Hobson and Friston (2012, 2014). However, in my view, a satisfactory account of dreams in this framework must start from their *conscious* (and affective) character.
26. Nour and Carhart-Harris (2017).
27. Dehaene and Changeux (2011), Friston, Breakspear and Deco (2012).
28. Montague et al. (2012), Corlett and Fletcher (2014), Friston, Stephan, Montague and Dolan (2014), Wang and Krystal (2014).
29. For a discussion of 'empathy' see Solms (2021b). The original German term is *Einfühlung*, which literally means 'feeling into'.

Chapter 10: Back to the Cortex

1. Zeki (1993), p. 236, emphasis added.
2. Ibid., p. 238.
3. I do not mean to imply that this is a high-level cognitive process. Visual 'projection' cortex appears to fill in the blind spot with whatever surrounds it. In natural conditions this is abetted by frequent eye movements which ensure that you almost always have an actual (ultra-short-term-memory) image of what is in the blind spot. For the variety of mechanisms involved, see Ramachandran (1992), Ramachandran and Gregory (1991) and Ramachandran, Gregory and Aiken (1993). Incidentally, alongside filling-in of the blind spot there is a filtering-out of unwanted objects, like the shadows of entoptic 'floaters' and your retinal vasculature.
4. The overlap between the fields represented by each eye cancels out the blind spots, which means you must close one of them to observe the illusory filling-in just mentioned.
5. There is, of course, some overlap between them, but the extent to which the objects of our visual attention literally are bisected is easily demonstrated in cases of unilateral (left) neglect following right-hemisphere damage.
6. This, incidentally, is evidence for the view that the 'binding' function of consciousness comes from below the cortex; from the unitary brainstem rather than the bicameral cortices (see Panksepp and Biven, 2012).
7. Solms et al. (1988).
8. He described it as 'a wonderful piece of theatre' (Helmholtz, 1867, p. 776).
9. Ibid., p. 438.
10. Most people working within the predictive coding paradigm do not pay

sufficient heed to the fact that perceptual inference is an *unconscious* process. For example, I do not agree with Hohwy (2013) when he says that 'what we are aware of is the "fantasy" generated by the way current predictions attenuate prediction error'. In my view, we do not *become aware* of our predictive 'fantasies' unless they *clash* with reality. And thank God for that.

It seems that Clark agrees with me on this score (Lupyan and Clark, 2015, p. 281, emphasis added):

> While most of the predictions are unconscious, one can sometimes become aware of them *when they are violated*. For example, imagine drinking from a glass of what you think is orange juice only to realize on tasting it that it is actually milk. The difference between the taste of that milk when one expects it and when one expects orange juice instead is the orange-juice expectation made conscious (Lupyan, 2015, for discussion). Similarly, consider the experience of an unexpected omission, as when a musical note is missing from a familiar composition. Such omissions can be as perceptually striking and as salient as the most vibrant tone – an otherwise puzzling effect that is neatly explained by assuming that the construction of perceptual experience involves expectations based upon some kind of model of what is likely to occur.

Despite what Hohwy (2013) says, he sometimes comes close to my view. In an earlier article he made the following statement (Hohwy, 2012, p. 11, emphasis added):

> This temporal signature is consistent with predictive coding insofar as when the prediction error from a stimulus is comprehensively suppressed and no further exploration is happening (since active inference is subdued due to central fixation during covert attention) probability should begin to drop. This follows from the idea that *what drives conscious perception is the actual process of suppressing prediction error*.

However, in his book (2013, p. 201, emphasis added) he says:

> Conscious perception is the upshot of unconscious perceptual inference. We are not consciously engaging in Bayesian updating of our priors in the light of new evidence, nor of the way sensory input is predicted and then attenuated. What is conscious is the result of the inference – *the conclusion*.

That is why I do not agree with Hohwy (2013), if I read him correctly. Where we agree is on the view that it is the *work* of minimising precise error signals that generates consciousness, i.e. the problematic mismatch between prediction and precision-weighted error. But I disagree with him when he says that 'what we are aware of is the "fantasy"'. At best, Hohwy seems to think that we are aware of the 'fantasy' that is trying to *explain away* the incoming error, but I think we are aware of the fact that it is *not* suppressing the error – i.e. we are aware of the predictive 'work in progress' caused by the mismatch. That is what makes reality salient. Perhaps this is just a semantic point. The bottom line for me is that consciousness in perception is driven by *uncertainty*, not by the *best-guesses* that relative certainty gives rise to. To my way of thinking, the guesses become conscious only when they are uncertain, and they recede when they are confirmed. In other words, we only become conscious of our fantasies when they are contradicted by reality. Consciousness might be described as a process of *disambiguation*.

I am aware that this issue is complicated by the difference between perceptual and active inference (and between exogenous and endogenous attention). This explains why there is a large body of literature – the literature upon which Hohwy (2013) relies – that suggests we preferentially perceive (consciously) that which conforms to our expectations *and* there is another body of literature which suggests that we preferentially perceive (consciously) that which is most unexpected. For me, this contradiction is resolved by the concept of *affect prioritisation*, which is driven, in both cases, by the balance between what I have loosely called 'needs' and 'opportunities' (see Chapter 5). In short, we perceive consciously that which is most salient *in relation to our currently prioritised need*. Prioritised needs (affects) yield the most precise error signals, by definition.

My views on this issue are elaborated further in the next section. Ultimately, however, this is an empirical question. Together with my students Donne van der Westhuizen and Julianne Blignaut and my ex-student Joshua Martin, I am currently researching the question using the standard binocular rivalry paradigm, which is a paradigm concerning perceptual *consciousness*. See also Pezzulo (2014), Yang, Zald and Blake (2007) and Stein and Sterzer (2012).

11. I first published this idea in Solms (2013) and then followed it up in various ways, e.g. Solms (2015b, 2017b, 2017c, 2018b).

12. Freud (1920), p. 25, emphasis added.

13. Freud assigned consciousness and memory to two different systems of neurons (ω and ψ, respectively), which later became his metapsychological systems *Cs.* and *Pcs.* – but he construed them *both* as cortical systems. The passage from which the sentence 'consciousness

arises instead of a memory-trace' is quoted makes this very clear, even though it was written twenty-five years after the 'Project' (see Freud, 1920, pp. 25 ff.).

14. Bargh and Chartrand (1999), p. 476:

Some of the automatic guidance systems we've outlined are 'natural' and don't require experience to develop. These are the fraternization of perceptual and behavioural representations and the connection between automatic evaluation processes on the one hand and mood and behaviour on the other. Other forms of automatic self-regulation develop out of repeated and consistent experience; they map onto the regularities of one's experience and take tasks over from conscious choice and guidance when that choice is not really being exercised. This is how goals and motives can come to operate non-consciously in given situations, how stereotypes can become chronically associated with the perceptual features of social groups, and how evaluations can become integrated with the perceptual representation of the person, object, or event so that they become active immediately and unintentionally in the course of perception.

15. Technically, neuronal states, which are boosted by neuronal precisions, update neuronal parameters.

16. This coincides exactly with Freud's notion of 'secondary process'. He conceptualised secondary process as a 'binding' of the primary form of drive energy, which is 'freely mobile'. The binding of Friston free energy is the mechanical foundation of what we call (effective) mental *work*. This conclusion is highly significant: 'In my opinion this distinction [between bound and free energy] represents the deepest insight we have gained up to the present into the nature of nervous energy, and I do not see how we can avoid making it' (Freud, 1915b, p. 188).

   Note, however, Freud took the view that the secondary process can also function *preconsciously*. This raises an empirical question: can cortical processing perform its stabilising function in the absence of consciousness? This question evokes the current controversy concerning non-declarative working memory (see Hassin et al., 2009).

17. I do not mean to imply that all reconsolidation entails consciousness *of the memory that is being updated*; it can entail affective modulation of the updating process which remains unconscious, cognitively (see below). Incidentally, the mechanisms I have just reviewed explain the psychological phenomenon that Freud called 'resistance': our peculiar

reluctance to update our predictive models in the face of contradictory evidence. (Unfortunately, this applies even to scientists!)

18. The term 'Reward = prediction' in Schultz's diagram might confuse some readers. When the sensory consequences of an action match the predicted consequences, nothing happens; there is no 'reward'. Feelings, both negative and positive, always signify error (see Figure 12). In behaviourist parlance, however, 'reward' does not imply any feeling. It simply means that the prediction is reinforced. That is, using my terminology, it is assigned greater precision. To me, that means it will be felt as pleasurable *if the relevant prediction was prioritised by the midbrain decision triangle.*

19. Misanin, Miller and Lewis (1968).

20. Nader, Schafe and LeDoux (2000). See Dudai (2000) for an accessible overview. Reconsolidation is closely related to Freud's concept of memory 'retranscription'. See his letter to Fliess dated 6 December 1896, in which he presages the concept of systems consolidation, at the least:

> As you know, I am working on the assumption that our psychical mechanism has come into being by a process of stratification: the material present in the form of memory traces being subjected from time to time to a *rearrangement* in accordance with fresh circumstances – to a *retranscription.* Thus what is essentially new about my theory is the thesis that memory is present not once but several times over, that it is laid down in various kinds of indications. I postulated a similar kind of rearrangement some time ago (*Aphasia* [Freud, 1891]) for the paths leading from the periphery. I cannot say how many of these registrations there are: at least three, probably more. This is shown in the following schematic picture, [where consciousness appears as the final 'registration' of the trace; see Figure 8] which assumes that the different registrations are also separated (not necessarily topographically) according to the neurons which are their vehicles. This assumption may not be necessary, but it is the simplest and is admissible provisionally.

> Freud conceptualised 'repression' as a failure of retranscription. See Solms (2017c) for a neuropsychoanalytic update.

21. Hence both long-term potentiation and long-term depression are modulated by the reticular activating system (see Bienenstock, Cooper and Munro, 1982). Hence also the capacity of ECT and epileptic seizures, both acting via the reticular activating system, to interfere with memory consolidation.

22. What Hebb (1949), p. 62, actually said was:

> Let us assume that the persistence or repetition of a reverberatory activity (or 'trace') tends to induce lasting cellular changes that add to its stability [...] When an axon of cell A is near enough to excite a cell B and repeatedly or persistently takes part in firing it, some growth process or metabolic change takes place in one or both cells such that A's efficiency, as one of the cells firing B, is increased.

23. Together with Cristina Alberini, I am currently planning a series of experiments on the roles of the PAG and reticular activating system in conscious vs unconscious learning. These experiments will clarify the roles of the different 'arousal' nuclei in the upper brainstem that modulate conscious cognition in the forebrain. The reconsolidation paradigm holds the promise of revealing some of the elementary intracellular mechanisms of perceptual consciousness in relation to learning. For example, both consolidation and reconsolidation can be disrupted by protein synthesis inhibition and both require the gene transcription factor CREB. However, recent research suggests that in the amygdala BDNF is required for consolidation but not reconsolidation and that the transcription factor and immediate early gene Zif268 is required for reconsolidation but not consolidation. A similar double dissociation between Zif268 for reconsolidation and BDNF for consolidation was found in the hippocampus. See Debiec et al. (2006) and Lee, Everitt and Thomas (2004).

24. Riggs and Ratliff (1951), Ditchburn and Ginsborg (1952). It is instructive to recall Helmholtz's observations on attention (1867, p. 770):

> The natural unforced state of our attention is to wander around to ever new things, so that when the interest of an object is exhausted, when we cannot perceive anything new, then attention against our will goes to something else [...] If we want attention to stick to an object we have to keep finding something new in it, especially if other strong sensations seek to decouple it.

   This should be linked with what I said about default SEEKING above.

25. See https://en.wikipedia.org/wiki/Lilac_chaser.

26. If anyone tells you these effects do not implicate the brain (i.e. that they occur only at more peripheral levels of the nervous system) tell them to read Hsieh and Tse (2006). See also Coren and Porac (1974).

27. See Oberauer et al. (2013).

28. I am excluding 'priming' and perceptual learning here. These do entail images and involve the cortex. Most of the other things I say about

non-declarative memory do not apply to priming and perceptual learning, which belong in a category of their own; they are the 'scaffolding' of declarative memory. See what I say further on about the priming function of words, for example.

29. The basal ganglia are spared in some (but not all) hydranencephalic children and decorticate animals.

30. Technically, complexity is the relative entropy between posterior and prior beliefs or probability distributions over external states. This definition of complexity follows from the fact that model evidence is the difference between accuracy and complexity (see Chapter 8). As model evidence is actively increased by minimising free energy, the accuracy of predictions rises, with a concomitant increase in complexity. In other words, increasing model complexity is always licensed by an ability to make more accurate predictions – as typically occurs in cortical memory systems.

31. Compare Hohwy (2013), p. 202:

> The idea would be that perceptual inference moves around in a space determined by both prediction error accuracy and prediction error precisions. This can be depicted in a simplified way, if we conceive of prediction error accuracy as increasing with the inverse amplitude of the prediction error itself, and prediction error precision as increasing with the inverse amplitude of random fluctuations around uncertainty about predictions [...] This points to a unified account of the relation between conscious perception and attention. They stand to each other as first order and second order statistical inference.

> See also Hohwy (2012), discussed briefly above.

32. This section relies heavily – and gratefully – upon Clark (2015).

33. The difference pivots on the fact that dreams are almost devoid of exteroceptive error signals, owing to the dramatic shifts in precision weighting that occur with sleep onset (Hobson and Friston, 2012, 2014). Clark describes this as 'insulation from entrainment'; but see note 35 below.

34. Clark (2015), p. 273.

35. However, note that the sleeping brain is not 'insulated from entrainment', as Clark puts it (2015), p. 107. The main 'driving sensory signal' of the brain is always endogenous. It is simply not possible to become insulated from entrainment by this signal – i.e. from our biological needs. If we neuroscientists continue to overlook this fundamental fact, we will never understand mental life and its place in nature.

36. See Domhoff (2017).

37. Clark (2015), p. 274.

38. See Solms (2021a) for an elaboration of this point.

39. This occurs also in the default-mode 'resting state'.

40. It is easy to forget that the hippocampus is part of the limbic system – the emotional brain.

41. Okuda et al. (2003), Szpunar et al. (2007), Szpunar (2010), Addis et al. (2007).

42. Ingvar (1985).

43. Schacter, Addis and Buckner (2007), p. 660, emphasis added.

44. The fact that a *changing* (unpredictable) stream suppresses a *constant* (more predictable) image is, of course, interesting in itself.

45. Lupyan and Ward (2013), p. 14196, emphasis added. In terms of the technical issues concerning conscious perception that were discussed above, this is, of course, an instance of endogenous attention. In this connection, see also the technical definition of 'salience' provided on p. 205. Words artificially boost salience. However, verbal priming is easily overridden by a strong (i.e. precise) exogenous surprise. In other words, bottom-up prioritisation of needs at the level of the midbrain decision triangle invariably trumps top-down forebrain processes.

46. Lupyan and Thompson-Schill (2012).

47. Çukur et al. (2013). See again the inattentional blindness experiment cited in Chapter 9, note 8.

48. Clark (2015), p. 286.

49. Roepstorff and Frith (2004).

50. See Zhou et al. (2017).

51. Panksepp and Biven (2012), p. 396:

    The *experience* of conscious sight and sound were initially largely affective (Panksepp, 1998). The immediacy with which sudden visual and auditory stimuli can startle and frighten us, especially when such stimuli originate very close to our bodies, suggests a deep primal integration of these sensory systems with some of our most essential affective survival mechanisms. Consider also how we are prone to associate certain colors with feelings.

52. Consider also the increasingly influential concept of 'affective touch'.

53. I am not saying that perceptual consciousness *as a whole* has no causal power. It acquires its power precisely because it contextualises affect.

54. See Hurley, Dennett and Adams (2011), a book about humour which Dennett recently brought to my notice because it comes to conclusions

about the functional architecture of the mind that are remarkably similar to my own.

55. Auditory and visual wavelengths and intensities are continuously measured and compared and classified by the cortex both consciously and unconsciously.

56. Cf. Clark (2015), p. 207: The free energy paradigm suggests 'not that we experience our own prediction error signals (or their associated precisions) as such. Instead, those signals act within us to recruit the apt flows of predictions that reveal a world of distal objects and causes.'

57. If you are alarmed by my use of the word 'avatar', remember that everything you perceive is virtual, including your image of your own body. Consider what I once wrote about the 'body swap' illusion (Solms, 2013, p. 15):

> The subject of consciousness identifies itself with its external body (object-presentation) in much the same way as a child projects itself into the animated figure it controls in a computer game. The representation is rapidly invested with a sense of self, although it is not really the self. Here is a striking experiment that vividly illustrates the counterintuitive relation that actually exists between the subjective self and its external body. Petkova and Ehrsson (2008) report a series of 'body swap' experiments in which cameras mounted over the eyes of other people, or mannequins, transmitting images from that viewpoint to video-monitoring goggles mounted over the eyes of the experimental subjects, rapidly created the illusion in the experimental subjects that the other person's body or the mannequin was their own body. This illusion was so compelling that it persisted even when the projected subjects shook hands with their own bodies. The existence of the illusion was also demonstrated objectively by the fact that when the other (illusory own) body and the (real) own body were both threatened with a knife, the fear response – the 'gut reaction' of the internal body (measured by heart rate and galvanic skin response) – was greater for the illusory body [...] We are reminded that cortex is nothing but random-access memory.

> On this last point, see Ellis and Solms (2018).

Chapter 11: The Hard Problem

1. Davies (2019), pp. 184, 207.
2. Crick (1994). The full quotation is given on p. 240.
3. Chalmers (1995a), p. 201, emphasis added.

4. Crick (1994), p. 3, emphasis added.

5. See Chalmers (1996), p. 251, emphasis added: 'Whoever would have thought that this hunk of gray matter would be the sort of thing that could *produce* vivid subjective experiences? And yet it does.'

6. Searle (1997), p. 28. Searle also puts it like this: 'How exactly do neurobiological processes in the brain *cause* consciousness?' (1993, p. 3, emphasis added).

7. I am referring here to the causal closure of the physical: if consciousness is not physical, or conscious properties are not physical properties, then it's hard to see how they can influence the causal matrix of brain processes. See René Descartes's correspondence with Princess Elisabeth of Bohemia (Shapiro, 2007).

8. Levine (1983). The irreducibility of phenomenal experience to physical processes is also known as the 'epistemic gap'.

9. Jackson (1982). I say 'something like this' because I have slightly modified Jackson's 'knowledge argument'. I have done so not only to simplify it, but also because it is unnecessarily cruel in its original form. In the real world, such cruelty would impact upon the psychological processes Jackson describes.

10. Chalmers (2003), p. 104.

11. See Locke's *Essay Concerning Human Understanding* (1690): 'It is impossible to conceive that matter, either with or without motion, could have originally, in and from itself, sense, perception, and knowledge; as is evident from hence, that then sense, perception, and knowledge must be a property eternally inseparable from matter and every particle of it.'

12. I do not mean to imply that Locke was an epiphenomenalist. There are, of course, other dualist positions, but Jackson (1982), who coined the thought experiment about Mary, adopted an epiphenomenal position. Later (1995) he changed his mind.

13. Oakley and Halligan (2017). These authors do at least attribute *some* function to consciousness: the ability to *report* on mental states (which are unconscious in themselves).

14. To paraphrase Chalmers, it is logically conceivable that 'philosophical zombies' could emulate all the mechanical functions of the brain without having any conscious experiences.

15. Libet et al. (1983). However, Libet himself does not believe that consciousness is epiphenomenal. He takes the view that during the 300 milliseconds leading up to an action, consciousness could choose to abort that action ('free won't'). This type of brain wave is called the

'readiness potential'. Subsequent research has suggested the latency period between the readiness potential and the conscious decision may be substantially more than 300 ms.

16. This, in turn, may be explained by Hebb's Law, which, for that very reason, also explains your lived experience of ten-minute-old memories not being so well consolidated. All of this is consistent, I should point out, with what Chalmers (1995a) calls the Principle of Structural Coherence; see p. 255.

17. Atkinson and Shiffrin (1971).

18. Short-term memory traces decay rapidly as a consequence of neurotransmitter reuptake mechanisms that restore presynaptic neurons to the state that existed prior to the formation of each trace; thereby enabling them rapidly to form further traces. See Mongillo, Barak and Tsodyks (2008).

19. Chalmers (1995a), pp. 202–3. Chalmers explains his use of the term 'function': 'Here "function" is not used in the narrow teleological sense of something that a system is designed to do, but in the broader sense of any causal role in the production of behaviour that a system might perform.'

20. Ibid., pp. 204–5.

21. Ibid., p. 205.

22. Ibid., p. 204.

23. Cf. Nagel (1974):

> If physicalism is to be defended, the phenomenological features [of consciousness] must themselves be given a physical account. But when we examine their subjective character it seems that such a result is impossible. The reason is that every subjective phenomenon is essentially connected with a single point of view, and it seems inevitable that an objective, physical theory will abandon that point of view.

Searle (1997), p. 212, puts it like this:

> Consciousness has a first person or subjective ontology and so cannot be reduced to anything that has third-person or objective ontology. If you try to reduce or eliminate one in favour of the other you leave something out […] You can neither reduce the neuron firings to the feelings nor the feelings to the neuron firings, because in each case you would leave out the objectivity or subjectivity that is in question.

24. Chalmers (1995a), p. 203.

25. Havlík, Kozáková and Horáče (2017).

26. Letter of 23 December 2017. Incidentally, we submitted our article to

the *Journal of Consciousness Studies* because that is where Chalmers (1995a) first formulated the hard problem.

27. Zahavi (2017) makes this point – ironically, against Friston's work.

28. I am well aware of Searle's objection here, to the effect that qualia present a case where a single reality cannot have multiple appearances, because they *are* the appearances (see note 39, below). My answer to him would be: 'Yes, but do not forget that the *visualised* appearance of somatosensory cortical activity and the *felt* appearance of pain are both part of that same experienced reality.'

29. In this analogy, I am referring to two *exteroceptive* phenomena: perceptions of lightning and thunder. When I ask what causes them both, I could address the question in two ways. Either I could describe the geophysical (electrical) events that generate them both or the sensory mechanisms that register (these two different aspects of) the physical events. I have chosen to do the former in this analogy and relegate the latter description to my consideration of the actual problem I have addressed in this book, namely the relationship between objective and subjective events. This is because, as we saw earlier, the *consciousness* attaching to sensory events, whether they be exteroceptively or interoceptively aroused, is always endogenous. This point cannot be made in the analogy. Consciousness itself, I am arguing, is not a sensory signal (exteroceptive or interoceptive) but rather the *feeling* of the signal.

30. You can do so using optogenetics, for example. Using different equipment, you can *listen* to spike trains of retinal signals. As Wheeler says, it is all just a matter of 'equipment-evoked responses'.

31. Chalmers (1995a) for this and what follows, emphasis added passim.

32. Chalmers's position appears at first sight to be conventional property dualism, in which mind is a property of matter. He then construes both mind and matter as properties of something else, called 'information' (see below). This might sound like dual-aspect monism but his is the curious type of information I discussed in previous chapters: the mind inheres in the information rather than in the receiver of it. The same applies to the matter aspect of information, as Chalmers has it. This means that the two aspects of information are not properties in the epistemological sense (they are not *appearances* of something called information) but rather in the ontological sense (they are aspects of information *itself*). Nevertheless, this is not the issue that should concern us. What concerns me most is Chalmers's claim that *all* information has a mental (indeed, a conscious) aspect to it. Whether he takes the mental aspect of information to be a property or a substance, therefore,

is not the main issue. The main issue becomes: is it plausible to attribute consciousness to all information?

33. Remember that, for Chalmers, 'the phenomenal features of the world' exclude 'the physical'.

34. I should say: the equipment *that is* the participant observer.

35. I got this analogy from Hurley, Dennett and Adams (2011).

36. This relates to what I said above (in Chapter 9) about intentional selfhood.

37. Chalmers (1995a), p. 217. See also his earlier remark: 'This leads to a natural hypothesis: that information (*or at least some information*) has two basic aspects, a physical aspect and a phenomenal aspect' (emphasis added).

38. See Chapter 10, note 14.

39. Cf. Searle (1992), p. 121–122: 'We can't [distinguish appearance from reality] for consciousness because consciousness consists in the appearances themselves. Where appearance is concerned we cannot make the appearance-reality distinction because the appearance is the reality.'

40. Incidentally, if Mary were an affective neuroscientist rather than a visual one, she could not be devoid of affective experiences in the same way as she was of visual ones. This is because, if she did not feel like something, she would be in a coma (if not dead).

41. See Chapter 5, note 4.

42. Ambulation by means of wheels is conceivable, but what actually evolved (in our case) was legs. We must be careful not to set a higher bar for explanations of consciousness than for everything else in biology.

43. Chalmers (1995a), pp. 203–4.

Chapter 12: Making a Mind

1. But an octopus might disagree. (The vertical lobe system seems to be the closest cephalopod analogue for the vertebrate pallium.)

2. This group consisted of Tristan Hromnik, Jonathan Shock and me, to begin with, and then gradually expanded to include other physicists, computer scientists and biomedical engineers – George Ellis, Rowan Hodson, Leen Remmelzwaal, Amit Mishra, Dean Rance, Dawie van den Heever and Julianne Blignaut – as well as neuropsychologists Joshua Martin, Aimee Dollman and Donne van der Westhuizen. The team continues to grow, although Hromnik is no longer part of it and Martin has moved to Berlin.

3. Searle (1980). Damasio (2018) took a similar view.

4. I am referring to the 'dancing qualia' argument of Chalmers, which

builds on Locke's 'inverted spectra' argument. See Chalmers (1995a,b, 2011).

5. Chalmers (1995a), pp. 214–15.

6. Ibid., p. 215, emphasis added.

7. Ibid., emphasis added.

8. I have had lengthy discussions with him on this score and my impression is that he has an open mind, at least.

9. Ethier et al. (2012), Hochberg et al. (2012), Collinger et al. (2013), Bouton et al. (2016), Capogrosso et al. (2016).

10. Capogrosso et al. (2016), p. 284.

11. Ibid.

12. Abu-Hassan et al. (2019).

13. Pasley et al. (2012).

14. Nishimoto et al. (2011).

15. Horikawa et al. (2013).

16. Herff et al. (2015).

17. Including top-down ones of the kind discussed under the heading of 'thinking' in Chapter 10 (such as imagining and dreaming).

18. Cf. Kurzweil (2005).

19. Solms and Turnbull (2002), pp. 70–71, second emphasis added.

20. See Solms (1996, 1997b).

21. The renowned physicist Richard Feynman took the same view about mechanistic understanding in general. I do not mean that reverse-engineering consciousness in and of itself solves the problem. It is possible to assemble something mindlessly microchip-by-microchip without understanding it. What I mean is that *if* you understand it, then you should be able to reverse-engineer it.

22. See Reggia (2013) for a review of previous research along these lines.

23. Since the reinforcement-learning approach requires a goal criterion, this will be it: *survival of the system in unpredicted environments.*

24. To be clear: embodiment can be simulated (as can all the physical parameters described below). From the viewpoint of the system, it doesn't matter what is really happening 'outside', only what is happening in the model in relation to *the information it receives* from outside. The research team can therefore simulate an environment for the system to model, and we will do so for the early generations of our proposed system. To proceed otherwise would be extremely time-consuming and frankly dangerous (consider the overheating parameter described below, for example). When I say that it is more 'realistic' to embody the system

physically, what I mean is that there are modelling problems that arise with physical movement (for example) that do not arise with simulated movement, and this might very well prove to be important for a truly lifelike system. For this reason, and others, the later generations of our proposed system will be embodied in robots.

25. I.e. we will encode artificial reflexes and instincts. We will also use 'genetic algorithms'.

26. This from Wikipedia (https://en.wikipedia.org/wiki/Artificial_consciousness) on 21 March 2020. For an alternative view see Reggia (2013):

> The author of this review believes that none of the past studies examined, even when claimed otherwise, has yet provided a convincing argument for how the approach being studied would eventually lead to instantiated artificial consciousness. On the other hand, and more positively, no evidence has yet been presented (including by the work surveyed in this review) that instantiated machine consciousness could not one day ultimately be possible, a view that has been expressed by others.

27. Solms and Turnbull (2002), pp. 68–9.

28. See Colby, Watt and Gilbert (1966), Weizenbaum (1976). A computer program called 'Eugene Goostman', which simulates a thirteen-year-old Ukrainian boy, passed the Turing test at an event held in 2014 at the Royal Society of London.

29. For example, as is well known, computers can outperform the very best human players of both chess and Go, which is more difficult than chess.

30. See Haikonen (2012).

31. This is touching, considering how much prejudice Alan Turing himself (the designer of the test) suffered.

32. Several years ago, I said at a meeting held in Vienna between psychoanalysts and AI engineers that one way of demonstrating artificial consciousness is to look for evidence of artificial psychopathology: 'To the extent that the engineers succeed in accurately emulating the human mind, to that extent they will find that their model is prone to certain types of malfunctioning. One is almost tempted to use this as a criterion of their success' (Solms, 2008).

33. Mathur, Lau and Guo (2011).

34. Consider the extreme case of 'decentralised autonomous corporations'.

35. See Lin, Abney and Bekey (2011).

36. Bill Gates, Stephen Hawking and Elon Musk, for example, have all expressed serious reservations on this topic.

37. I do not claim that this belief alone provides ethical justification for doing it. The fact that someone will commit murder somewhere, someday, does not justify my committing murder here and now. Please read on …

38. Solms and Friston (2018), Solms (2019a), Solms (2020).

39. Here is a partial list: 'Where does consciousness fit in the Bayesian brain?' 18th International Neuropsychoanalysis Congress, University College London, 2017; 'How and why consciousness arises', Department of Physics, University of Cape Town 2017; 'How and why consciousness arises', The Centre for Subjectivity Research, University of Copenhagen, 2017; 'The self as feeling and memory', Ruhr University, Bochum, 2018; 'The conscious id, the psychoanalytic process and the hard problem of consciousness', Department of Philosophy, New York University, 2019; 'Why and how consciousness arises', Department of Psychiatry, Mount Sinai Hospital, New York, 2019; 'Why are we conscious? Lessons from neuroscience', University of Vermont College of Medicine, Burlington, 2019; 'What is consciousness?' The Melbourne Brain Centre, Australia, 2019; 'Consciousness itself', The Science of Consciousness, Interlaken, Switzerland, 2019; 'Consciousness itself is affect', Munich School of Philosophy, Burkardus Haus, Würzburg, Germany, 2019; 'The hard problem of consciousness', Department of Philosophy, University of Cape Town, 2019; 'Consciousness is predictive work in progress', Ichilov Hospital, Tel Aviv, 2019; 'Why and how consciousness arises', Italian Psychoanalytic Dialogues, Rome, 2020.

40. Havlík, Kozáková and Horáče (2017).

41. Carhart-Harris and Friston (2010), Carhart-Harris et al. (2014), Carhart-Harris (2018).

42. See Chapter 9, note 16.

43. www.linking-ai-principles.org/term/656. See The Public Voice Coalition (2018).

Postscript

1. The Law of Affect: 'If a behaviour is consistently accompanied by pleasure it will increase, and if it is consistently accompanied by unpleasure it will decrease.'

2. Friston's Law: 'All the quantities that can change, i.e. that are part of the system, will change to minimise free energy.'

References

Abbott, A. (2020), What animals really think. *Nature*, 584: 182–5

Absher, J. and Benson, D. (1993), Disconnection syndromes: an overview of Geschwind's contributions. *Neurology*, 43: 862–7

Abu-Hassan, K., Taylor, J., Morris, P. et al. (2019), Optimal solid state neurons. *Nature Communications*, 10: 5309

Adams, R., Shipp, S. and Friston, K. (2013), Predictions not commands: active inference in the motor system. *Brain Structure and Function*, 218: 611–43

Addis, D., Wong, A. and Schacter, D. (2007), Remembering the past and imagining the future: common and distinct neural substrates during event construction and elaboration. *Neuropsychologia*, 45: 1363–77

Ainley, V., Apps, M. A. J., Fotopoulou, A. and Tsakiris, M. (2016), 'Bodily precision': a predictive coding account of individual differences in interoceptive accuracy. *Philosophical Transactions of the Royal Society of London, B*, 371: 2016003 doi.org/10.1098/rstb.2016.0003

Alboni, P. and Alboni, M. (2014), Vasovagal syncope as a manifestation of an evolutionary selected trait. *Journal of Atrial Fibrillation*, 7: 1035

Aserinsky, E. and Kleitman, N. (1953), Regularly occurring periods of eye motility, and concomitant phenomena, during sleep. *Science*, 118: 273–4

Ashby, W. (1947), Principles of the self-organizing dynamic system. *Journal of General Psychology*, 37: 125–8

Atkinson, R. and Shiffrin, R. (1971), The control of short-term memory. *Scientific American*, 225: 82–90

Baars, B. (1988), *A Cognitive Theory of Consciousness*. Cambridge: Cambridge University Press

Baars, B. (1997), *In the Theatre of Consciousness*. Oxford: Oxford University Press

Bailey, P. and Davis, E. (1942), The syndrome of obstinate progression in the cat. *Experimental Biology and Medicine*, 52: 307

Bargh, J. and Chartrand, T. (1999), The unbearable automaticity of being. *American Psychologist*, 54: 462–79

Barrett, L. F. (2017), *How Emotions are Made: The Secret Life of the Brain*. New York: Houghton Mifflin Harcourt

Bastos, A., Usrey, W., Adams, R. et al. (2012), Canonical microcircuits for predictive coding. *Neuron*, 76: 695–711

Bastos, A., Vezoli, J., Bosman, C. et al. (2015), Visual areas exert feedforward and feedback influences through distinct frequency channels. *Neuron*, 85: 390–401

Bayes, T. (1763), An essay towards solving a problem in the doctrine of chances. [Communicated by Mr. Price, in a letter to John Canton.] *Philosophical Transactions of the Royal Society of London*, 53: 370–418

Bechtel, W. and Richardson, R. (1998), Vitalism. In E. Craig (ed.), *Routledge Encyclopedia of Philosophy*, 9. London: Routledge, pp. 639–43

Bentley, B., Branicky, R., Barnes, C. et al. (2016), The multilayer connectome of *Caenorhabditis elegans*. *PLoS Computational Biology*, 12: e1005283, doi.org/10.1371/journal.pcbi.1005283

Berlin, H. (2011), The neural basis of the dynamic unconscious. *Neuropsychoanalysis*, 13: 5–31

Berlin, H. (2013), The brainstem begs the question: 'petitio principii'. *Neuropsychoanalysis*, 15: 25–9

Berridge, K. (2003), Pleasures of the brain. *Brain and Cognition*, 52: 106–28

Besharati, S., Forkel, S. J., Kopelman, M., Solms, M., Jenkinson, P. M. and Fotopoulou, A. (2014), The affective modulation of motor awareness in anosognosia for hemiplegia: behavioural and lesion evidence. *Cortex*, 61: 127–40

Besharati, S., Forkel, S., Kopelman, M., Solms, M., Jenkinson, P. and Fotopoulou, A. (2016), Mentalizing the body: spatial and social cognition in anosognosia for hemiplegia. *Brain*, 139: 971–85

Besharati, S., Fotopoulou, A. and Kopelman, M. (2014), What is it like to be confabulating? In A. L. Mishara, A. Kranjec, P, Corlett, P. Fletcher and M. A. Schwartz (eds.), *Phenomenological Neuropsychiatry, How Patient Experience Bridges Clinic with Clinical Neuroscience*. New York: Springer

Bienenstock, E., Cooper L. and Munro P. (1982), Theory for the development of neuron selectivity: orientation specificity and binocular interaction in visual cortex. *Journal of Neuroscience*, 2: 32–48

Blake, Y., Terburg, D., Balchin, R., van Honk, J. and Solms, M. (2019), The role of the basolateral amygdala in dreaming, *Cortex*, 113: 169–83, doi.org/10.1016/j.cortex.2018.12.016

Block, N. (1995), On a confusion about a function of consciousness. *Behavioral and Brain Sciences*, 18: 227–47

Blomstedt, P., Hariz, M., Lees, A. et al. (2008), Acute severe depression induced by intraoperative stimulation of the substantia nigra: a case report. *Parkinsonism and Related Disorders*, 14: 253–6

Bogen, J. (1995), On the neurophysiology of consciousness: 1. An overview. *Consciousness and Cognition*, 4: 52–62

Bouton, C., Shaikhouni, A., Annetta, N. et al. (2016), Restoring cortical control of functional movement in a human with quadriplegia. *Nature*, 533: 247–50

Bowlby, J. (1969), *Attachment*. London: Hogarth Press

Braun, A. (1999), The new neuropsychology of sleep. *Neuropsychoanalysis*, 1: 196–201

Braun, A., Balkin, T., Wesenten, N. et al. (1997), Regional cerebral blood flow throughout the sleep-wake cycle. An H2(15)O PET study. *Brain*, 120: 1173–97

Brentano, F. (1874), *Psychologie vom empirischen Standpunkte*. Leipzig: Duncker and Humbolt

Broca, P. (1861), Sur le principe des localisations cérébrales. *Bulletin de la Société d'Anthropologie*, 2: 190–204

Broca, P. (1865), Sur le siège de la faculté du langage articulé. *Bulletin de la Société d'Anthropologie*, 6: 377–93

Brown, H., Adams, R., Parees, I., Edwards, M. and Friston, K. (2013), Active inference, sensory attenuation and illusions. *Cognitive Processing*, 14: 411–27

Cameron-Dow, C. (2012), Do dreams protect sleep? Testing the Freudian hypothesis of the function of dreams. MA dissertation, University of Cape Town

Campbell, A. (1904), Histological studies on the localisation of cerebral function. *Journal of Mental Science*, 50: 651–62

Capogrosso, M., Milekovic, T., Borton, D. et al. (2016), A brain–spine interface alleviating gait deficits after spinal cord injury in primates. *Nature*, 539: 284–8

Carhart-Harris, R. (2018), The entropic brain – revisited. *Neuropharmacology*, 142: 167–78

Carhart-Harris, R. and Friston, K. (2010), The default-mode, ego-functions and free-energy: a neurobiological account of Freudian ideas. *Brain*, 133: 1265–83

Carhart-Harris, R., Leech, R., Hellyer, P. et al. (2014), The entropic brain:

a theory of conscious states informed by neuroimaging research with psychedelic drugs. *Frontiers in Human Neuroscience*, 8: Article 20

Chabris, C. and Simons, D. (2010), *The Invisible Gorilla: and Other Ways Our Intuitions Deceive Us*. London: Crown Publishers/Random House

Chalmers, D. (1995a), Facing up to the problem of consciousness. *Journal of Consciousness Studies*, 2: 200–219

Chalmers, D. (1995b), Absent qualia, fading qualia, dancing qualia. In T. Metzinger (ed.), *Conscious Experience*. Paderborn: Ferdinand Schoningh, pp. 309–28

Chalmers, D. (1996), *The Conscious Mind: In Search of a Fundamental Theory*. New York: Oxford University Press

Chalmers, D. (2003), Consciousness and its place in nature. In S. Stich and T. Warfield (eds.), *Blackwell Guide to the Philosophy of Mind*. London: Blackwell, pp. 102–42

Chalmers, D. (2011), A computational foundation for the study of cognition. *Journal of Cognitive Science*, 12: 325–59

Charcot J-M. (1883), Un cas de suppression brusque et isolée de la vision mentale des signes et des objets (formes et couleurs). *Progrès Médical*, 11: 568–71

Chew, Y., Tanizawa, Y., Cho, Y. et al. (2018), An afferent neuropeptide system transmits mechanosensory signals triggering sensitization and arousal in *C. elegans*. *Neuron*, 99: 1233–46

Cisek, P. and Kalaska, J. (2010), Neural mechanisms for interacting with a world full of action choices. *Annual Review of Neuroscience*, 33: 269–98

Claparède, E. (1911), Recognition et moitié. *Archives de psychologie*, 11: 79–90

Clark, A. (2015), *Surfing Uncertainty: Prediction, Action, and the Embodied Mind*. New York: Oxford University Press

Clark, A. (2017), Busting out: predictive brains, embodied minds, and the puzzle of the evidentiary veil. *Noûs*, 51: 727–53

Coenen, A. (2007), Consciousness without a cortex, but what kind of consciousness is this? *Behavioral and Brain Sciences*, 30: 87–8

Coenen, V., Bewernick, B., Kayser, S. et al. (2019), Superolateral medial forebrain bundle deep brain stimulation in major depression: a gateway trial. *Neuropsychopharmacology*, 44: 1224–32, doi.org/10.1038/s41386-019-0369-9

Colby, K., Watt, J. and Gilbert, J. (1966), A computer method of psychotherapy. *Journal of Nervous and Mental Disease*, 142: 148–52

Cole, S., Fotopoulou, A., Oddy, M. and Moulin, C. (2014), Implausible future events in a confabulating patient with an anterior communicating artery aneurysm. *Neurocase*, 20: 208–24

Collinger J., Wodlinger, B., Downey, J. et al. (2013), High-performance neuroprosthetic control by an individual with tetraplegia. *The Lancet*, 381: 557–64

Coltheart, M. and Turner, M. (2009), Confabulation and delusion. In W. Hirstein (ed.), *Confabulation: Views from Neuroscience, Psychiatry, Psychology and Philosophy*. New York: Oxford University Press, pp.173–88

Conant, R. and Ashby, W. (1970), Every good regulator of a system must be a model of that system. *International Journal of Systems Science*, 1: 89–97

Coren, S. and Porac, C. (1974), The fading of stabilized images: Eye movements and information processing. *Perception & Psychophysics*, 16: 529–34

Corlett, P. and Fletcher, P. (2014), Computational psychiatry: a Rosetta Stone linking the brain to mental illness. *Lancet Psychiatry*, 1: 399–402

Craig, A. D. (2009), How do you feel – now? The anterior insula and human awareness. *Nature Reviews Neuroscience*, 10: 59–70

Craig, A. D. (2011), Significance of the insula for the evolution of human awareness of feelings from the body. *Annals of the New York Academy of Sciences*, 1225: 72–82

Crick, F. (1994), *The Astonishing Hypothesis: The Scientific Search for the Soul*. New York: Charles Scribner's Sons

Crick, F. (2004), Foreword to C. Koch, *The Quest for Consciousness: A Neurobiological Approach*. Englewood, CO: Roberts and Company

Crick, F. and Koch, C. (1990), Towards a neurobiological theory of consciousness. *Seminars in the Neuroscience*, 2: 263–75

Crucianelli, L., Krahé, C., Jenkinson, P. and Fotopoulou, A. (2017), Interoceptive ingredients of body ownership: affective touch and cardiac awareness in the rubber hand illusion. *Cortex*, 104: 180–92, doi.org/10.1016/j.cortex.2017.04.018

Çukur, T., Nishimoto, S., Huth, A. and Gallant J. (2013), Attention during natural vision warps semantic representation across the human brain. *Nature Neuroscience*, 16: 763–70

Dahan, L., Astier, B., Vautrelle, N. et al. (2007), Prominent Burst Firing of Dopaminergic Neurons in the Ventral Tegmental Area during Paradoxical Sleep. *Neuropsychopharmacology*, 32: 1232–41

Damasio, A. (1994), *Descartes' Error: Emotion, Reason, and the Human Brain*. New York: Putnam

Damasio, A. (2018), *The Strange Order of Things: Life, Feeling, and the Making of Cultures*. London: Penguin Random House

Damasio, A. and Carvalho, G. (2013), The nature of feelings: evolutionary and neurobiological origins. *Nature Reviews Neuroscience*, 14: 143–52

Damasio, A. and Damasio, H. (1989), *Lesion Analysis in Neuropsychology*. New York: Oxford University Press

Damasio, A., Damasio, H. and Tranel, D. (2013), Persistence of feelings and sentience after bilateral damage of the insula. *Cerebral Cortex*, 23: 833–46

Damasio, A., Grabowski, T., Bechara, A. et al. (2000), Subcortical and cortical brain activity during the feeling of self-generated emotions. *Nature Neuroscience*, 3: 1049–56

Darwin, C. (1859), *On the Origin of Species*. London: John Murray

Darwin, C. (1872), *The Expression of Emotions in Man and Animals*. London: John Murray

Davies, P. (2019), *The Demon in the Machine: How Hidden Webs of Information are Solving the Mystery of Life*. London: Allen Lane

Debiec, J., Doyere, V., Nader, K. and LeDoux, J. (2006), Directly reactivated, but not indirectly reactivated, memories undergo reconsolidation in the amygdala. *Proceedings of the National Academy of Sciences*, 103: 3428–33

Decety, J. and Fotopoulou, A. (2015), Why empathy has a beneficial impact on others in medicine: unifying theories. *Frontiers in Behavioral Neuroscience*, 8: 457

Dehaene, S. and Changeux, J.-P. (2005), Ongoing spontaneous activity controls access to consciousness: a neuronal model for inattentional blindness. *PLoS Biology*, 3: e141

Dehaene, S. and Changeux, J.-P. (2011), Experimental and theoretical approaches to conscious processing. *Neuron*, 70: 200–227

Dehaene, S. and Naccache, L. (2001), Towards a cognitive neuroscience of consciousness: basic evidence and a workspace framework. *Cognition*, 79: 1–37

Dement, W. and Kleitman, N. (1957), The relation of eye movements during sleep to dream activity: an objective method for the study of dreaming. *Journal of Experimental Psychology*, 53: 339–46

Depaulis, A. and Bandler, R. (1991), *The Midbrain Periaqueductal Gray Matter: Functional, Anatomical, and Neurochemical Organization*. New York: Plenum Press

Ditchburn, R. and Ginsborg, B. (1952), Vision with a stabilized retinal image, *Nature*, 170: 36–7

Domhoff, W. (2017), *The Emergence of Dreaming: Mind-Wandering, Embodied Simulation, and the Default Network*. New York: Oxford University Press

Du Bois-Reymond, E. (1848–84), *Untersuchungen über thierische Electricität*, 2. Berlin: Reimer

Du Bois-Reymond, E., ed. (1918), *Jugendbriefe von Emil Du Bois-Reymond an Eduard Hallmann, zu seinem hundertsten Geburtstag, dem 7. November 1918*. Berlin: Reimer

Dudai, Y. (2000), The shaky trace. *Nature*, 406: 686–7

Edelman, G. (1990), *The Remembered Present: A Biological Theory of Consciousness*. New York: Basic Books

Edlow, B., Takahashi, E., Wu, O. et al. (2012), Neuroanatomic connectivity of the human ascending arousal system critical to consciousness and its disorders. *Journal of Neuropathology and Experimental Neurology*, 71: 531–46

Einstein, A. (1905), Über einen die Erzeugung und Verwandlung des Lichtes betreffenden heuristischen Gesichtspunkt. *Annalen der Physik*, 17: 132–48

Eisenberger, N. (2012), The neural bases of social pain: evidence for shared representations with physical pain. *Psychosomatic Medicine*, 74: 126–35

Ekman, P., Friesen, W., O'Sullivan, M. et al. (1987), Universals and cultural differences in the judgements of facial expressions of emotion. *Journal of Personality and Social Psychology*, 53: 712–17

Ellis, G. and Solms, M. (2018), *Beyond Evolutionary Psychology: How and Why Neuropsychological Modules Arise*. Cambridge: Cambridge University Press

England, J. (2013), Statistical physics of self-replication. *Journal of Chemical Physics*, 139: 121923, doi.org/10.1063/1.4818538

Ethier, C., Oby, E., Bauman, M. and Miller, L. (2012), Restoration of grasp following paralysis through brain-controlled stimulation of muscles. *Nature*, 485: 368–71

Ezra, M., Faull, O., Jbabdi, S. and Pattinson, K. (2015), Connectivity-based segmentation of the periaqueductal gray matter in human with brainstem optimized diffusion MRI. *Human Brain Mapping*, 36: 3459–71

Feldman, H. and Friston, K. J. (2010), Attention, uncertainty, and free-energy. *Frontiers in Human Neuroscience*, 4: 215, doi.org/10.3389/fnhum.2010.00215

Ferrarelli, F. and Tononi, G. (2011), The thalamic reticular nucleus and schizophrenia. *Schizophrenia Bulletin*, 37: 306–15

Fischer, D., Boes, A., Demertzi, A. et al. (2016), A human brain network derived from coma-causing brainstem lesions. *Neurology*, 87: 2427–34

Flechsig, P. (1901), Developmental (mylogenetic) localisation of the cerebral cortex in the human subject. *The Lancet*, 2: 1027–9

Flechsig, P. (1905), Gehirnphsyiologie und Willenstheorien. *Fifth International Psychology Congress*, Rome, pp. 73–89. In G. von Bonin (ed.), *Some Papers on the Cerebral Cortex*. Springfield, IL: Charles C. Thomas, pp. 181–200

Forrester, G., Davis, R., Mareschal, D. et al. (2018), The left cradling bias: an evolutionary facilitator of social cognition? *Cortex*, 118: 116–31, doi.org/10.1016/j.cortex.2018.05.011

Fotopoulou, A. (2008), False-selves in neuropsychological rehabilitation: the challenge of confabulation. *Neuropsychological Rehabilitation*, 18: 541–65

Fotopoulou, A. (2009), Disentangling the motivational theories of confabulation. In W. Histein (ed.), *Confabulation: Views from Neurology, Psychiatry, and Philosophy*. New York: Oxford University Press

Fotopoulou, A. (2010a), The affective neuropsychology of confabulation and delusion. *Cognitive Neuropsychiatry*, 15: 38–63

Fotopoulou, A. (2010b), The affective neuropsychology of confabulation and delusion. In R. Langdon and M. Turner (eds.), *Confabulation and Delusion*, New York: Psychology Press, pp. 38–63

Fotopoulou, A. (2013), Beyond the reward principle: consciousness as precision seeking. *Neuropsychoanalysis*, 15: 33–8

Fotopoulou, A. and Conway, M. (2004), Confabulation pleasant and unpleasant. *Neuropsychoanalysis*, 6: 26–33

Fotopoulou, A., Conway, M., Birchall, D., Griffiths, P. and Tyrer, S. (2007), Confabulation: revising the motivational hypothesis. *Neurocase*, 13: 6–15

Fotopoulou, A., Conway, M. and Solms, M. (2007), Confabulation: motivated reality monitoring. *Neuropsychologia*, 45: 2180–90

Fotopoulou, A., Conway, M., Solms, M., Tyrer, S. and Kopelman, M. (2008a), Self-serving confabulation in prose recall. *Neuropsychologia*, 46: 1429–41

Fotopoulou, A., Conway, M., Tyrer, S., Birchall, D., Griffiths, P. and Solms, M. (2008b), Is the content of confabulation positive? An experimental study. *Cortex*, 44: 764–72.

Fotopoulou, A., Solms, M. and Turnbull, O. (2004), Wishful reality distortions in confabulation: a case report. *Neuropsychologia*, 42: 727–44

Fotopoulou, A. and Tsakiris, M. (2017), Mentalizing homeostasis: the social origins of interoceptive inference. *Neuropsychoanalysis*, 19: 3–76

Frank, J. (1946), Clinical survey and results of 200 cases of prefrontal leucotomy. *Journal of Mental Sciences*, 92: 497–508

Frank, J. (1950), Some aspects of lobotomy (prefrontal leucotomy) under psychoanalytic scrutiny. *Psychiatry*, 13: 35–42

Frank, M. (2005), Dynamic dopamine modulation in the basal ganglia: A neurocomputational account of cognitive deficits in medicated and nonmedicated Parkinsonism. *Journal of Cognitive Neuroscience*, 17: 51–72

Freud, S. (1883), Einleitung in der Nervenpathologie. Unpublished book manuscript. Washington, DC: Library of Congress

Freud, S. (1886), Report on my studies in Paris and Berlin. *Standard Edition of the Complete Psychological Works of Sigmund Freud*, 1. London: Hogarth, pp. 1–15

Freud, S. (1888), Gehirn. I. Anatomie des Gehirns. In A. Villaret (ed.), *Handwörterbuch der gesamten Medizin*, 1. Stuttgart: Ferdinand Enke, pp. 684–91

Freud, S. (1891), *On Aphasia*. New York: International Universities Press

Freud, S. (1893a), Charcot. *Standard Edition of the Complete Psychological Works of Sigmund Freud*, 3. London: Hogarth, pp. 11–23

Freud, S. (1893b), Some points for a comparative study of organic and hysterical motor paralyses. *Standard Edition of the Complete Psychological Works of Sigmund Freud*, 1. London: Hogarth, pp. 155–72

Freud, S. (1894), The neuro-psychoses of defence. *Standard Edition of the Complete Psychological Works of Sigmund Freud*, 3. London: Hogarth, pp. 45–61

Freud, S. (1895), Studies on hysteria. *Standard Edition of the Complete Psychological Works of Sigmund Freud*, 2. London: Hogarth

Freud, S. (1900), The interpretation of dreams. *Standard Edition of the Complete Psychological Works of Sigmund Freud*, 4 and 5. London: Hogarth

Freud, S. (1901), The psychopathology of everyday life. *Standard Edition of the Complete Psychological Works of Sigmund Freud*, 6. London: Hogarth

Freud, S. (1912), A note on the unconscious in psycho-analysis. *Standard*

Edition of the Complete Psychological Works of Sigmund Freud, 12. London: Hogarth, pp. 255–66

Freud, S. (1914), On narcissism: an introduction. *Standard Edition of the Complete Psychological Works of Sigmund Freud*, 14. London: Hogarth, pp. 67–102

Freud, S. (1915a), Instincts and their vicissitudes. *Standard Edition of the Complete Psychological Works of Sigmund Freud*, 14. London: Hogarth, pp. 117–40

Freud, S. (1915b), The unconscious. *Standard Edition of the Complete Psychological Works of Sigmund Freud*, 14. London: Hogarth, pp. 166–204

Freud, S. (1920), Beyond the pleasure principle. *Standard Edition of the Complete Psychological Works of Sigmund Freud*, 18. London: Hogarth, pp. 7–64

Freud, S. (1923), The ego and the id. *Standard Edition of the Complete Psychological Works of Sigmund Freud*, 19. London: Hogarth, pp. 12–59

Freud, S. (1925), A note upon 'the mystic writing-pad'. *Standard Edition of the Complete Psychological Works of Sigmund Freud*, 19. London: Hogarth, pp. 227–32

Freud, S. (1940 [1939]), An outline of psycho-analysis. *Standard Edition of the Complete Psychological Works of Sigmund Freud*, 23. London: Hogarth, pp. 144–207

Freud, S. (1950a [1895]), Extracts from the Fliess papers. *Standard Edition of the Complete Psychological Works of Sigmund Freud*, 1. London: Hogarth, pp. 177–280

Freud, S. (1950b [1895]), Project for a scientific psychology. *Standard Edition of the Complete Psychological Works of Sigmund Freud*, 1. London: Hogarth, pp. 283–397.

Freud, S. (1994 [1929]), Letter to Einstein, 1929. In I. Grubrich-Simitis (1995), 'No greater, richer, more mysterious subject … than the life of the mind'. *International Journal of Psychoanalysis*, 76: 115–22

Friston, K. (2005), A theory of cortical responses. *Philosophical Transactions of the Royal Society of London, B*, 360: 815–36

Friston, K. (2009), The Free Energy Principle: a rough guide to the brain? *Trends in Cognitive Sciences*, 13: 293–301

Friston, K. (2013), Life as we know it. *Journal of the Royal Society Interface*, 10: 20130475, doi.org/10.1098/rsif.2013.0475

Friston, K., Breakspear, M. and Deco, G. (2012), Perception and self-organized instability. *Frontiers in Computational Neuroscience*, 6: 44

Friston, K., Rigoli, F., Ognibene, D. et al. (2015), Active inference and epistemic value. *Cognitive Neuroscience*, 6: 187–214

Friston, K., Schwartenbeck, P., FitzGerald, T., Moutoussis, M., Behrens, T. and Dolan, R. (2014), The anatomy of choice: dopamine and decision-making. *Philosophical Transactions of the Royal Society of London, B*, 369: doi.org/10.1098/rstb.2013.0481

Friston, K., Shiner, T., FitzGerald, T., Galea, J., Adams, R., Brown, H., Dolan, R., Moran, R., Stephan, K. and Bestmann, S. (2012), Dopamine, affordance and active inference. *PLoS Computational Biology*, 8: e1002327

Friston, K. and Stephan, K. (2007), Free-energy and the brain. *Synthese*, 159: 417–58

Friston, K., Stephan, K., Montague, R. and Dolan, R. (2014), Computational psychiatry: the brain as a phantastic organ. *Lancet Psychiatry*, 1: 148–58

Frith, C., Blakemore, S. and Wolpert, D. (2000), Abnormalities in the awareness and control of action. *Philosophical Transactions of the Royal Society of London, B*, 355: 1771–88

Galin, D. (1974), Implications for psychiatry of left and right cerebral specialization: a neurophysiological context for unconscious processes. *Archives of General Psychiatry*, 31: 572–83

Garcia-Rill, E. (2017), Bottom-up gamma and stages of waking. *Medical Hypotheses*, 104: 58–62

Gloor, P. (1992), Role of the amygdala in temporal lobe epilepsy. In J. Aggleton (ed.), *The Amygdala: Neurobiological Aspects of Emotion, Memory, and Mental Dysfunction*. New York: Wiley-Liss, pp. 505–38

Golaszewski, S. (2016), Coma-causing brainstem lesions. *Neurology*, 87: 10

Goodglass, H. (1986), Norman Geschwind (1926–1984). *Cortex*, 22: 7–10

Gosseries, O., Schnakers, C., Ledoux, D. et al. (2011), Automated EEG entropy measurements in coma, vegetative state/unresponsive wakefulness syndrome and minimally conscious state. *Functional Neurology*, 26: 25–30

Gregory, R. (1980), Perceptions as hypotheses. *Philosophical Transactions of the Royal Society of London, B*, 290: 181–97

Haikonen, P. (2012), *Consciousness and Robot Sentience*. New Jersey: World Scientific

Harding, D. (1961), *On Having No Head*. London: Sholland Trust

Harlow, J. (1868), Passage of an iron rod through the head. *Boston Medical and Surgical Journal*, 39: 389–93

Hartmann. E., Russ. D., Oldfield. M., Falke. R. and Skoff. B. (1980), Dream content: effects of ı-DOPA. *Sleep Research*, 9: 153

Hassin, R., Bargh, J., Engell, A. and McCulloch, K. (2009), Implicit working memory. *Consciousness and Cognition*, 18: 665–78

Havlík, M., Kozáková, E. and Horáče, J. (2017), Why and how: the future of the central questions of consciousness. *Frontiers in Psychology*, 8: 1797, doi.org/10.3389/fpsyg.2017.01797

Hebb, D. (1949), *The Organization of Behavior: A Neuropsychological Theory*. New York: Wiley

Helmholtz, H. von (1867), *Handbuch der physiologischen Optik*, 3. Leipzig: Voss

Helmholtz, H. von (1892), Goethes Vorahnungen kommender naturwissenschaftlicher Ideen. In *Vorträge und Reden*, 2. Braunschweig: Friedrich Vieweg und Sohn, pp. 335–61

Herff, C., Heger, D., de Pesters, A. et al. (2015), Brain-to-text: decoding spoken phrases from phone representations in the brain. *Frontiers in Neuroscience*, 9: 217, doi.org/10.3389/fnins.2015.00217

Hering, E. (1879), Der Raumsinn und die Bewegungen des Auges. In L. Hermann (ed.), *Handbuch der Physiologie*, 3. Part 1: *Physiologie des Gesichtssinnes*. Leipzig: Vogel, pp. 343–601

Hesselmann, G., Sadaghiani, S., Friston, K. and Kleinschmidt, A. (2010), Predictive coding or evidence accumulation? False inference and neuronal fluctuations. *PLoS One*, 5(3): e9926, doi.org/10.1371/journal.pone.0009926

Hobson, J. A. (2009), REM sleep and dreaming: towards a theory of protoconsciousness. *Nature Reviews Neuroscience*, 10: 803–13

Hobson, J. A. and Friston (2012), Waking and dreaming consciousness: neurobiological and functional considerations. *Progress in Neurobiology*, 98: 82–98

Hobson, J. A. and Friston, K. (2014), Consciousness, dreams, and inference: the Cartesian theatre revisited. *Journal of Consciousness Studies*, 21: 6–32

Hobson, J. A. and McCarley, R. (1977), The brain as a dream state generator: an activation-synthesis hypothesis of the dream process. *American Journal of Psychiatry*, 134: 1335–48

Hobson, J. A., McCarley, R. and Wyzinski, P. (1975), Sleep cycle oscillation: reciprocal discharge by two brainstem neuronal groups. *Science*, 189: 55–8

Hochberg L., Bacher, D., Jarosiewicz, B. et al. (2012), Reach and grasp by people with tetraplegia using a neurally controlled robotic arm. *Nature*, 485: 372–5

Hohwy, J. (2012), Attention and conscious perception in the hypothesis testing brain. *Frontiers in Psychology*, 3: 96, doi.org/10.3389/fpsyg.2012.00096

Hohwy, J. (2013), *The Predictive Mind*. New York: Oxford University Press

Holeckova, I., Fischer, C., Giard, M.-H. et al. (2006), Brain responses to subject's own name uttered by a familiar voice. *Brain Research*, 1082: 142–52

Holstege, G., Georgiadis, J., Paans, A. et al. (2003), Brain activation during human male ejaculation. *Journal of Neuroscience*, 23: 9185–93

Horikawa, T., Tamaki, M., Miyawaki, Y. and Kamitani, Y. (2013), Neural decoding of visual imagery during sleep. *Science*, 340: 639–42

Hsieh, P-J. and Tse, P. (2006), Illusory color mixing upon perceptual fading and filling-in does not result in 'forbidden colors'. *Vision Research*, 46: 2251–8

Hume, D. (1748), *Philosophical Essays Concerning Human Understanding*. London: A. Millar.

Hurley, M., Dennett, D. and Adams, R. (2011), *Inside Jokes: Using Humor to Reverse-Engineer the Mind*. Cambridge, MA: MIT Press

Ingvar, D. (1985), 'Memory of the future': an essay on the temporal organization of conscious awareness. *Human Neurobiology*, 4: 127–36

Jackson, F. (1982), Epiphenomenal qualia. *Philosophical Quarterly*, 32: 127–36

Jackson, F. (1995), Postscript on 'What Mary Didn't Know'. In P. Moser and J. Trout (eds.), *Contemporary Materialism*. London: Routledge, pp. 184–9

Jaspers, K. (1963), *General Psychopathology*. Chicago: University of Chicago Press

Jaynes, E. (1957), Information theory and statistical mechanics. *Physical Review*, 106: 620–30

Jouvet, M. (1965), Paradoxical sleep: a study of its nature and mechanisms. *Progress in Brain Research*, 18: 20–62

Joyce, J. (2008), Bayes' theorem. *Stanford Encyclopedia of Philosophy*

Julesz, B. (1971), *Foundations of Cyclopean Perception*. Chicago: University of Chicago Press

Kandel, E. (1998), A new intellectual framework for psychiatry. *American Journal of Psychiatry*, 155: 457–69

Kandel, E. (1999), Biology and the future of psychoanalysis: a new intellectual framework for psychiatry revisited. *American Journal of Psychiatry*, 156: 505–24

Kant, I. (1790), Kritik der Urteilskraft. *Kants gesammelte Schriften*, 5. Berlin: Walter de Gruyter

Kaplan-Solms, K. and Solms, M. (2000), *Clinical Studies in Neuro-Psychoanalysis: Introduction to a Depth Neuropsychology*. London: Karnac

Kihlstrom, J. (1996), Perception without awareness of what is perceived, learning without awareness of what is learned. In M. Velmans (ed.), *The Science of Consciousness: Psychological, Neuropsychological and Clinical Reviews*. London: Routledge, pp. 23–46

Knill, J. and Pouget, A. (2004), The Bayesian brain: the role of uncertainty in neural coding and computation. *Trends in Neurosciences*, 27: 712–19

Koch, C. (2004), *The Quest for Consciousness: A Neurobiological Approach*. Englewood, CO: Roberts and Company

Kopelman, M., Bajo, A. and Fotopoulou, A. (2015), Confabulation: memory deficits and neuroscientific aspects. In J. Wright (ed.), *International Encyclopedia of Social and Behavioral Sciences*. New York: Elsevier

Krahé, C., Springer, A., Weinman, J. and Fotopoulou, A. (2013), The social modulation of pain: others as predictive signals of salience – a systematic review. *Frontiers in Human Neuroscience*, 7: 386

Kurzweil, R. (2005), *The Singularity is Near: When Humans Transcend Biology*. New York: Viking

Lane, N. (2015), *The Vital Question: Why is Life the Way It Is?* London: Profile

Lavie, P., Pratt, H., Scharf, B., Peled, R. and Brown, J. (1984), Localized pontine lesion: nearly total absence of REM sleep. *Neurology*, 34: 118–20

LeDoux, J. (1996), *The Emotional Brain*. New York: Simon and Schuster

LeDoux, J. (1999), Psychoanalytic theory: clues from the brain. *Neuropsychoanalysis*, 1: 44–9

LeDoux, J. and Brown, R. (2017), A higher-order theory of emotional consciousness. *Proceedings of the National Academy of Science*, 114: e2016–e2025

Lee, J., Everitt, B. and Thomas, K. (2004), Independent cellular processes for hippocampal memory consolidation and reconsolidation. *Science*, 304: 839–43

Léna, I., Parrot, S., Deschaux, O. et al. (2005), Variations in extracellular levels of dopamine, noradrenaline, glutamate, and aspartate across the sleep–wake cycle in the medial prefrontal cortex and nucleus accumbens of freely moving rats. *Journal of Neuroscience Research*, 81: 891–9

Leng, G. (2018), *The Heart of the Brain: The Hypothalamus and Its Hormones*. Cambridge, MA: MIT Press

LeVay, S. (1993), *The Sexual Brain*. Cambridge, MA: MIT Press

Levine, J. (1983), Materialism and qualia: the explanatory gap. *Pacific Philosophical Quarterly*, 64: 354–61

Libet, B., Gleason, C., Wright E. and Pearl, D. (1983), Time of conscious intention to act in relation to onset of cerebral activity (readiness-potential): the unconscious initiation of a freely voluntary act. *Brain*, 106: 623–42

Lichtheim, L. (1885), On aphasia. *Brain*, 7: 433–84

Liepmann, H. (1900), Das Krankheitsbild der Apraxie ('motorischen Asymbolie') auf Grund eines Falles von einseitiger Apraxie. *Monatsschrift für Psychiatrie und Neurologie*, 8: 182–197

Lightman, A. (2018), *Searching for Stars on an Island in Maine*. New York: Pantheon

Lin, P., Abney, K. and Bekey, G. (eds.) (2011), *Robot Ethics*. Cambridge, MA: MIT Press

Linnman, C., Moulton, E., Barmettler, G. et al. (2012), Neuroimaging of the periaqueductal gray: state of the field. *Neuroimage*, 60: 505–22

Lisman, J. and Buzsaki, G. (2008), A neural coding scheme formed by the combined function of gamma and theta oscillations. *Schizophrenia Bulletin*, 34: 974–80

Lissauer, H. (1890), Ein Fall von Seelenblindheit, nebst einem Beitrag zur Theorie derselben. *Archiv für Psychiatrie und Nervenkrankheiten*, 21: 222–70

Lupyan, G. (2015), Cognitive penetrability of perception in the age of prediction: Predictive systems are penetrable systems. *Review of Philosophy and Psychology*, 6: 547–69

Lupyan, G. and Clark, A. (2015), Words and the world: predictive coding and the language-perception-cognition interface. *Current Directions in Psychological Science*, 24: 279–84

Lupyan, G. and Thompson-Schill, S. (2012), The evocative power of words: activation of concepts by verbal and nonverbal means. *Journal of Experimental Psychology – General*, 141: 170–86

Lupyan, G. and Ward, E. (2013), Language can boost otherwise unseen objects into visual awareness. *Proceedings of the National Academy of Sciences*, 110: 14196–201

Malcolm-Smith, S., Koopowitz, S., Pantelis, E. and Solms, M. (2012), Approach/avoidance in dreams. *Consciousness and Cognition*, 21: 408–12

Man, K. and Damasio, A. (2019), Homeostasis and soft robotics in the design of feeling machines. *Nature Machine Intelligence*, 1: 446–52, doi.org/10.1038/s42256–019–0103–7

Maturana, H. and Verela, F. (1972), *Autopoiesis and Cognition: The Realization of the Living*. London: Dordrecht

Marc, J. and Llinas, R. (1994), Human oscillatory brain activity near 40 Hz coexists with cognitive temporal binding. *Proceedings of the National Academy of Sciences*, 91: 11748–51

Mathur, P., Lau B. and Guo, S. (2011), Conditioned place preference behavior in zebrafish. *Nature Protocols*, 6: 338–45

Maxwell, J. (1872), *Theory of Heat*. London: Longmans, Green and Co.

Mazur, J. E. (2013), Basic principles of operant conditioning. *Learning and Behavior*, 7th edn. New York: Pearson, pp. 101–26

McCarley, R. and Hobson, J. A. (1977), The neurobiological origins of psychoanalytic dream theory. *American Journal of Psychiatry*, 134: 1211–21

McKeever, W. (1986), Tachistoscopic methods in neuropsychology. In H. J. Hannay (ed.), *Experimental Techniques in Human Neuropsychology*. Oxford: Oxford University Press, pp. 167–211

Merker, B. (2007), Consciousness without cerebral cortex: a challenge for neuroscience and medicine. *Behavioral and Brain Sciences*, 30: 63–81

Mesulam, M. M. (2000), Behavioral neuroanatomy: large-scale networks, association cortex, frontal syndromes, the limbic system, and hemispheric specializations. In *Principles of Behavioral and Cognitive Neurology*, 2nd edn. New York: Oxford University Press, pp. 1–120

Meynert, T. (1867), Der Bau der Gross-Hirnrinde und seine örtliche Verschiedenheiten, nebst einem pathologisch-anatomischen Corollarium. *Vierteljahrsschrift für Psychiatrie in ihren Beziehungen zur Morphologie und Pathologie des Central-Nervensystems, die physiologischen Psychologie, Statistik und gerichtlichen Medizin*, 1: 77–93, 119–24

Meynert, T. (1884), *Psychiatrie: Klinik der Erkrankungen des Vorderhirns*. Vienna: W. Braumüller

Misanin, J., Miller, R. and Lewis, D. (1968), Retrograde amnesia produced by electroconvulsive shock after reactivation of a consolidated memory trace. *Science*, 160: 554–5

Mohr, M. von and Fotopoulou, A. (2017), The cutaneous borders of interoception: active and social inference of pain and pleasure on the skin. In M. Tsakiris and H. de Preester (eds.), *The Interoceptive Basis of the Mind.* Oxford: Oxford University Press

Mongillo, G. Barak, O. and Tsodyks, M. (2008), Synaptic theory of working memory. *Science,* 319: 1543–6

Montague, P., Dolan R., Friston, K. and Dayan P. (2012), Computational psychiatry. *Trends in Cognitive Sciences,* 16: 72–80

Moruzzi, G. and Magoun, H. (1949), Brain stem reticular formation and activation of the EEG. *Electroencephalography and Clinical Neurophysiology,* 1: 455–73

Motta, S., Carobrez, A. and Canteras, N. (2017), The periaqueductal gray and primal emotional processing critical to influence complex defensive responses, fear learning and reward seeking. *Neuroscience and Biobehavioral Reviews,* 76: 39–47

Moustafa, A., Sherman, S. and Frank, M. (2008), A dopaminergic basis for working memory, learning and attentional shifting in Parkinsonism. *Neuropsychologia,* 46: 3144–56

Mulert, C., Menzinger, E., Leicht, G. et al. (2005), Evidence for a close relationship between conscious effort and anterior cingulate cortex activity. *International Journal of Psychophysiology,* 56: 65–80

Munk, H. (1878), Weiteres zur Physiologie des Sehsphäre der Grosshirnrinde. *Deutsche medizinische Wochenschrift,* 4: 533–6

Munk, H. (1881), *Über die Functionen der Grosshirnrinde: gesammelte Mittheilungen aus den Jahren 1877–80.* Berlin: Albrecht Hirschwald

Nader, K., Schafe, G. and LeDoux, J. (2000), Fear memories require protein synthesis in the amygdala for reconsolidation after retrieval. *Nature,* 406: 722–6

Nagel, T. (1974), What is it like to be a bat? *Philosophical Review,* 83: 435–50

Newman, J. and Baars, B. (1993), A neural attentional model for access to consciousness: a global workspace perspective. *Concepts in Neuroscience,* 4: 255–90

Niedenthal, P. (2007), Embodying emotion. *Science,* 316: 1002–5

Nishimoto, S., Vu, A., Naselaris, T. et al. (2011), Reconstructing visual experiences from brain activity evoked by natural movies. *Current Biology,* 21: 1641–6

Nour, M. and Carhart-Harris, R. (2017), Psychedelics and the science of self-experience. *British Journal of Psychiatry,* 210: 177–9

Nummenmaa, L., Hari, R., Hietanen, J. and Glerean, E. (2018), Maps of subjective feelings. *Proceedings of the National Academy of Sciences*, 115: 9198–203

Oakley, D. and Halligan, P. (2017), Chasing the rainbow: the non-conscious nature of being. *Frontiers in Psychology*, 8: 1924; doi.org/10.3389/fpsyg.2017.01924

Oberauer, K., Souza, A., Druey, M. and Gade, M. (2013), Analogous mechanisms of selection and updating in declarative and procedural working memory: experiments and a computational model. *Cognitive Psychology*, 66: 157–211

Okuda J., Fujii, T., Ohtake, H., Tsukiura, T., Tanji, K., Suzuki, K., Kawashima, R., Fukuda, H., Itoh, M. and Yamadori A. (2003), Thinking of the future and past: the roles of the frontal pole and the medial temporal lobes. *Neuroimage*, 19: 1369–80

Pace-Schott, E. and Hobson, J. A. (1998), Review of 'The Neuropsychology of Dreams'. *Trends in Cognitive Sciences*, 2: 199–200

Paloyelis, Y., Krahé, C., Maltezos, S., Williams, S., Howard, M. and Fotopoulou, A. (2016), The analgesic effect of oxytocin in humans: a double-blind, placebo-controlled cross-over study using laser-evoked potentials. *Journal of Neuroendocrinology*, 28: 10.1111/jne.12347

Panksepp, J. (1974), Hypothalamic regulation of energy balance and feeding behavior. *Federation Proceedings*, 33: 1150–65

Panksepp, J. (1998), *Affective Neuroscience: The Foundations of Human and Animal Emotions.* New York: Oxford University Press

Panksepp, J. (2011), The basic emotional circuits of mammalian brains: do animals have affective lives? *Neuroscience and Biobehavioral Reviews*, 35: 1791–804

Panksepp, J. and Biven, L. (2012), *The Archaeology of Mind: Neuroevolutionary Origins of Human Emotions.* New York: Norton

Panksepp, J. and Burgdorf, J. (2003), 'Laughing' rats and the evolutionary antecedents of human joy. *Physiology and Behavior*, 79: 533–47

Panksepp, J. and Solms, M. (2012), What is neuropsychoanalysis? Clinically relevant studies of the minded brain. *Trends in Cognitive Sciences*, 16: 6–8

Parr, T. and Friston, K. (2018), The anatomy of inference: generative models and brain structure. *Frontiers in Computational Neuroscience*, 12: 90, doi.org/10.3389/fncom.2018.00090

Partridge, M. (1950), *Pre-Frontal Leucotomy: A Survey of 300 Cases Personally Followed for 1½–3 Years.* Oxford: Blackwell

Parvizi, J. and Damasio, A. (2003), Neuroanatomical correlates of brainstem coma. *Brain*, 126: 1524–36

Pasley, B., David, S., Mesgarani, N. et al. (2012), Reconstructing speech from human auditory cortex. *PLoS Biology*, 10(1): e1001251, doi.org/10.1371/journal.pbio.1001251

Pellis S. and Pellis V. (2009), *The Playful Brain: Venturing to the Limits of Neuroscience*. Oxford: One World

Penfield, W. and Jasper, H. (1954), *Epilepsy and the Functional Anatomy of the Human Brain*. Little, Brown and Co.

Petkova, V. and Ehrsson, H. (2008), If I were you: perceptual illusion of body swapping. *PLoS One*, 3: e3832, doi.org/10.1371/journal.pone.0003832

Pezzulo, G. (2014), Why do you fear the bogeyman? An embodied predictive coding model of perceptual inference. *Cognitive Affective and Behavioral Neuroscience*, 14: 902–11

Pfaff, D. (2005), *Brain Arousal and Information Theory*. Cambridge, MA, Harvard University Press

Picard, F. and Friston K. (2014), Predictions, perception, and a sense of self. *Neurology*, 83: 1112–18

Popper, K. (1963), *Conjectures and Refutations*. London: Routledge

Qin, P., Di, H., Liu, Y. et al. (2010), Anterior cingulate activity and the self in disorders of consciousness. *Human Brain Mapping*, 31: 1993–2002

Ramachandran, V. S. (1992), Filling in the blind spot. *Nature*, 356: 115

Ramachandran, V. S. and Gregory, R. (1991), Perceptual filling in of artificially induced scotomas in human vision. *Nature*, 350: 699–702

Ramachandran, V. S., Gregory, R. and Aiken, W. (1993), Perceptual fading of visual texture borders. *Vision Research*, 33: 717–21

Reggia, J. (2013), The rise of machine consciousness: studying consciousness with computational models. *Neural Networks*, 44: 112–31

Riggs, L. and Ratliff, F. (1951), Visual acuity and the normal tremor of the eyes. *Science*, 114: 17–18

Roepstorff, A. and Frith, C. (2004), What's at the top in the top-down control of action? Script-sharing and 'top-top' control of action in cognitive experiments. *Psychological Research*, 68: 189–98

Rolls, E. (2014), *Emotion and Decision-Making Explained*. New York: Oxford University Press

Rolls, E. (2019), Emotion and reasoning in human decision-making. *Economics*, 13: 1–31

Rosenthal, D. (2005), *Consciousness and Mind*. Oxford: Oxford University Press

Rovelli, C. (2017), *Reality Is Not What It Seems: The Journey to Quantum Gravity*. Riverhead Books

Runes, D. (1972), *Dictionary of Philosophy*. Totowa, NJ: Littlefield, Adams and Co.

Sacks, O. (1970), *Migraine*. London: Vintage

Sacks, O. (1973), *Awakenings*. London: Duckworth

Sacks, O. (1984), *A Leg to Stand On*. New York: Simon and Schuster

Sacks, O. (1985), *The Man Who Mistook His Wife for a Hat*. London: Duckworth

Schacter, D., Addis, D. and Buckner, R. (2007), The prospective brain: remembering the past to imagine the future. *Nature Reviews Neuroscience*, 8: 657–61

Schindler, R. (1953), Das Traumleben der Leukotomierten. *Wiener Zeitschrift für Nervenheilkunde*, 6: 330

Searle, J. (1980), Minds, brains, and programs. *Behavioral and Brain Sciences*, 3: 417–24

Searle, J. (1992), *The Rediscovery of the Mind*. Cambridge, MA: MIT Press

Searle, J. (1993), The problem of consciousness. *Social Research*, 60: 3–16

Searle, J. (1997), *The Mystery of Consciousness*. London: Granta

Seeley, W., Menon, V., Schatzberg, A. et al. (2007), Dissociable intrinsic connectivity networks for salience processing and executive control. *Journal of Neuroscience*, 27: 2349–56

Seth, A. (2013), Interoceptive inference, emotion, and the embodied self. *Trends in Cognitive Sciences*, 17: 565–73

Shallice, T. (1988), *From Neuropsychology to Mental Structure*. Cambridge: Cambridge University Press

Shannon, C. (1948), A mathematical theory of communication. *The Bell System Technical Journal*, 27: 379–423

Shapiro, L. (2007), *The Correspondence Between Princess Elisabeth of Bohemia and René Descartes: Princess Elisabeth of Bohemia and René Descartes*. Chicago: University of Chicago Press

Sharf, B., Moskovitz, C., Lupton, M. and Klawans, H. (1978), Dream phenomena induced by chronic levodopa therapy. *Journal of Neural Transmission*, 43: 143–51

Shewmon, D., Holmes, G. and Byrne, P. (1999), Consciousness in congenitally decorticate children: developmental vegetative state as self-fulfilling prophecy. *Developmental Medicine and Child Neurology*, 41: 364–74

Skinner, B. F. (1953), *Science and Human Behavior*. New York: Macmillan

Solms, M. (1991), Summary and discussion of the paper 'The

neuropsychological organisation of dreaming: implications for psychoanalysis'. *Bulletin of the Anna Freud Centre*, 16: 149–65

Solms, M. (1995), New findings on the neurological organization of dreaming: implications for psychoanalysis. *Psychoanalytic Quarterly*, 64: 43–67

Solms, M. (1996), Was sind Affekte? *Psyche*, 50: 485–522

Solms, M. (1997a), *The Neuropsychology of Dreams: A Clinico-Anatomical Study*. Mahwah NJ: Lawrence Erlbaum Associates

Solms, M. (1997b), What is consciousness? *Journal of the American Psychoanalytic Association*, 45: 681–703

Solms, M. (1998), Before and after Freud's 'Project'. In R. Bilder and F. LeFever (eds.), Neuroscience of the Mind on the Centennial of Freud's Project for a Scientific Psychology. *Annals of the New York Academy of Sciences*, 843: 1–10

Solms M. (2000a), Dreaming and REM sleep are controlled by different brain mechanisms. *Behavioral and Brain Sciences*, 23: 843–50

Solms, M. (2000b), Freud, Luria and the clinical method. *Psychoanalysis and History*, 2: 76–109

Solms, M. (2000c), A psychoanalytic perspective on confabulation. *Neuropsychoanalysis*, 2: 133–8

Solms, M. (2001), The neurochemistry of dreaming: cholinergic and dopaminergic hypotheses. In E. Perry, H. Ashton and A. Young (eds.), *The Neurochemistry of Consciousness*. New York: John Benjamins, pp. 123–31

Solms, M. (2008), What is the 'mind'? A neuro-psychoanalytical approach. In D. Dietrich, G. Fodor, G. Zucker and D. Bruckner (eds.), *Simulating the Mind: A Technical Neuropsychoanalytical Approach*. Vienna: Springer Verlag, pp. 115–22

Solms, M. (2011), Neurobiology and the neurological basis of dreaming. In P. Montagna and S. Chokroverty (eds.), *Handbook of Clinical Neurology*, 98 (3rd series), *Sleep Disorders*, Part 1. New York: Elsevier, pp. 519–44

Solms, M. (2013), The conscious id. *Neuropsychoanalysis*, 15: 5–19

Solms, M. (2015a), *The Feeling Brain: Selected Papers on Neuropsychoanalysis*. London: Karnac

Solms, M. (2015b), Reconsolidation: turning consciousness into memory. *Behavioral and Brain Sciences*, 38, 40–41

Solms, M. (2017a), Empathy and other minds – a neuropsychoanalytic perspective and a clinical vignette. In V. Lux and S. Weigl (eds.),

  Empathy: Epistemic Problems and Cultural-Historical Perspectives of a Cross-Disciplinary Concept. London: Palgrave Macmillan, pp. 93–114

Solms, M. (2017b), Consciousness by surprise: a neuropsychoanalytic approach to the hard problem. In R. Poznanski, J. Tuszynski and T. Feinberg (eds.), *Biophysics of Consciousness: A Foundational Approach*. New York: World Scientific, pp. 129–48

Solms, M. (2017c), What is 'the unconscious', and where is it located in the brain? A neuropsychoanalytic perspective. *Annals of the New York Academy of Sciences*, 1406: 90–97

Solms, M. (2018a), Review of A. Damasio, 'The Strange Order of Things'. *Journal of the American Psychoanalytic Association*, 66: 579–86

Solms, M. (2018b), The scientific standing of psychoanalysis. *British Journal of Psychiatry – International*, 15: 5–8

Solms, M. (2019a), The hard problem of consciousness and the Free Energy Principle. *Frontiers in Psychology*, 9: 2714, doi.org/10.3389/fpsyg.2018.02714

Solms, M. (2019b), Commentary on Edmund Rolls: 'Emotion and reason in human decision-making'. *Economics Discussion Papers*, No. 2019–45, Kiel: Institute for the World Economy

Solms, M. (2020), New project for a scientific psychology: general scheme. *Neuropsychoanalysis*, 22: 1–31, doi.org/10.1080/15294145.2020.1833361

Solms, M. (2021a), Dreams and the hard problem of consciousness. In S. Della Salla (ed.), *Encyclopedia of Behavioral Neuroscience*. New York: Oxford University Press, in press

Solms, M. (2021b), Notes on some technical terms whose translation calls for comment. In M. Solms (ed.), *Revised Standard Edition of the Complete Psychological Works of Sigmund Freud*, 24. Lanham, MD: Rowman and Littlefield, in press

Solms, M. and Friston, K. (2018), How and why consciousness arises: some considerations from physics and physiology. *Journal of Consciousness Studies*, 25: 202–38

Solms, M., Kaplan-Solms, K. and Brown, J. W. (1996), Wilbrand's case of 'mind-blindness'. In C. Code, C.-W. Walesch, A.-R. Lecours and Y. Joanette (eds.), *Classic Cases in Neuropsychology*. Hove: Erlbaum, pp. 89–110

Solms, M., Kaplan-Solms, K., Saling, M. and Miller, P. (1988), Inverted vision after frontal lobe disease. *Cortex*, 24: 499–509

Solms, M. and Panksepp, J. (2010), Why depression feels bad. In E. Perry, D. Collerton, F. LeBeau and H. Ashton (eds.), *New Horizons in the Neuroscience of Consciousness*. Amsterdam: John Benjamins, pp. 169–79

Solms, M. and Saling, M. (1986), On psychoanalysis and neuroscience: Freud's attitude to the localizationist tradition. *International Journal of Psychoanalysis*, 67: 397–416

Solms, M. and Saling, M. (1990), *A Moment of Transition: Two Neuroscientific Articles by Sigmund Freud*. London: Karnac

Solms, M. and Turnbull, O. (2002), *The Brain and the Inner World: An Introduction to the Neuroscience of Subjective Experience*. London: Karnac

Solms, M. and Turnbull, O. (2011), What is neuropsychoanalysis? *Neuropsychoanalysis*, 13: 133–45

Solms, M. and Zellner, M. (2012), Freudian drive theory today. In A. Fotopoulou, D. Pfaff and M. Conway (eds.), *From the Couch to the Lab: Trends in Psychodynamic Neuroscience*. New York: Oxford University Press, pp. 49–63

Squire, L. (2009), The legacy of Patient HM for neuroscience. *Neuron*, 61: 6–9

Stein, T. and Sterzer, P. (2012), Not just another face in the crowd: detecting emotional schematic faces during continuous flash suppression. *Emotion*, 12: 988–96

Stoerig, P. and Barth, E. (2001), Low level phenomenal vision despite unilateral destruction of primary visual cortex. *Consciousness and Cognition*, 10: 574–87

Strawson, G. (2006), Realistic monism – why physicalism entails panpsychism. *Journal of Consciousness Studies*, 13: 3–31

Sulloway, F. (1979), *Freud: Biologist of the Mind*. New York: Burnett

Szpunar, K. (2010), Episodic future thought: an emerging concept. *Perspectives on Psychological Science*, 5: 142–62

Szpunar, K., Watson, J. and McDermott, K. (2007), Neural substrates of envisioning the future. *Proceedings of the National Academy of Sciences*, 104: 642–7

The Public Voice Coalition (2018), Universal guidelines for Artificial Intelligence, Draft 9, 23 October. Brussels: Electronic Privacy Information Center; https://thepublicvoice.org/ai-universal-guidelines

Thorndike, E. (1911), *Animal Intelligence*. New York: Macmillan

Tononi, G. (2012), Integrated information theory of consciousness: an updated account. *Archives of Italian Biology*, 150: 56–90

Tossani, E. (2013), The concept of mental pain. *Psychotherapy and Psychosomatics*, 82: 67–73

Tozzi, A., Zare, M. and Benasich, A. (2016), New perspectives on spontaneous brain activity: dynamic networks and energy matter. *Frontiers in Human Neuroscience*, 10: 247, doi.org/10.3389/fnhum.2016.00247

Tranel, D., Gullickson, G., Koch, M. and Adolphs, R. (2006), Altered experience of emotion following bilateral amygdala damage. *Cognitive Neuropsychiatry*, 11: 219–32

Turnbull, O. Berry, H. and Evans, C. (2004), A positive emotional bias in confabulatory false beliefs about place. *Brain and Cognition*, 55: 490–94

Turnbull, O., Bowman, C., Shanker, S. and Davies, J. (2014), Emotion-based learning: insights from the Iowa Gambling Task. *Frontiers in Psychology*, 5: 162, doi.org/10.3389/fpsyg.2014.00162

Turnbull, O., Fotopoulou, A. and Solms, M. (2014), Anosognosia as motivated unawareness: the 'defence' hypothesis revisited. *Cortex*, 61: 18–29

Turnbull, O., Jenkins, S. and Rowley, M. (2004), The pleasantness of false beliefs: an emotion-based account of confabulation. *Neuropsychoanalysis*, 6: 5–16

Turnbull, O. and Solms, M. (2007), Awareness, desire, and false beliefs. *Cortex*, 43: 1083–90

Uhlhaas, P. and Singer, W. (2010), Abnormal neural oscillations and synchrony in schizophrenia. *Nature Reviews Neuroscience*, 11: 100–113

van der Westhuizen, D., Moore, J., Solms, M., van Honk, J. (2017), Testosterone facilitates the sense of agency. *Consciousness and Cognition*, 56: 58–67

van der Westhuizen, D. and Solms, M. (2015), Social dominance and the Affective Neuroscience Personality Scales. *Consciousness and Cognition*, 33: 90–111

Varela, F., Thompson, E. and Rosch, E. (1991), *The Embodied Mind: Cognitive Science and Human Experience*. Cambridge, MA: MIT Press

Venkatraman, A., Edlow, B. and Immordino-Yang, M. (2017), The brainstem in emotion: a review. *Frontiers in Neuroanatomy*, 11: 15, doi.org/10.3389/fnana.2017.00015

Vertes, R. and Kocsis, B. (1997), Brainstem-diencephalo-septohippocampal systems controlling the theta rhythm of the hippocampus. *Neuroscience*, 81: 893–926

Walker, M. (2017), *Why We Sleep*. London: Penguin

Wang, X. and Krystal, J. (2014), Computational psychiatry. *Neuron*, 84: 638–54

Weiskrantz, L. (2009), *Blindsight: A Case Study Spanning 35 Years and New Developments*. New York: Oxford University Press

Weizenbaum, J. (1976), *Computer Power and Human Reason: From Judgment to Calculation*. New York: W. H. Freeman

Wernicke, C. (1874), *Der aphasische Symptomencomplex. Eine psychologische Studie auf anatomischer Basis*. Breslau: M. Crohn and Weigert

Wheeler, J. A. (1990). *Information, Physics, Quantum: The Search for Links*. In W. Zurek (Ed.), *Complexity, Entropy, and the Physics of Information*. Redwood City, CA: Addison-Wesley Publishing Company, pp. 310–1.

White, B., Berg, D., Kan, J. et al. (2017), Superior colliculus neurons encode a visual saliency map during free viewing of natural dynamic video. *Nature Communications*, 8: 14263, doi.org/10.1038/ncomms14263

Whitty, C. and Lewin, W. (1957), Vivid daydreaming; an unusual form of confusion following anterior cingulectomy. *Brain*, 80: 72–6

Wilbrand, H. (1887), *Die Seelenblindheit als Herderscheinung und ihre Beziehungen zur Homonymen Hemianopsie zur Alexie und Agraphie*. Wiesbaden: J. F. Bergmann

Wilbrand H. (1892), Ein Fall von Seelenblindheit und Hemianopsie mit Sectionsbefund. *Deutsche Zeitschrift für Nervenheilkunde*, 2: 361–87

Yang, E. Zald, D. and Blake, R. (2007), Fearful expressions gain preferential access to awareness during continuous flash suppression. *Emotion*, 7: 882–6

Yovell, Y., Bar, G., Mashiah, M. et al. (2016), Ultra-low-dose buprenorphine as a time-limited treatment for severe suicidal ideation: a randomized controlled trial. *American Journal of Psychiatry*, 173: 491–8

Yu, C. K.-C. (2007), Cessation of dreaming and ventromesial frontal-region infarcts. *Neuropsychoanalysis*, 9: 83–90

Zahavi, D. (2017), Brain, mind, world: predictive coding, neo-Kantianism, and transcendental idealism. *Husserl Studies*, 34: 47–61, doi.org/10.1007/s10743-017-9218-z

Zeki, S. (1993), *A Vision of the Brain*. Oxford: Blackwell

Zellner, M., Watt, D., Solms, M. and Panksepp, J. (2011), Affective neuroscientific and neuropsychoanalytic approaches to two intractable psychiatric problems: why depression feels so bad and what addicts really want. *Neuroscience and Biobehavioral Reviews*, 35: 2000–2008

Zeman, A. (2001), Consciousness. *Brain*, 124: 1263–89

Zhou, T., Zhu, H., Fan, Z. et al. (2017), History of winning remodels thalamo-PFC circuit to reinforce social dominance. *Science*, 357: 162–8

Index